LINEAR ALGEBRA
THROUGH ITS APPLICATIONS

LINEAR ALGEBRA

THROUGH ITS APPLICATIONS

T. J. FLETCHER

VAN NOSTRAND REINHOLD COMPANY
LONDON

NEW YORK   CINCINNATI   TORONTO   MELBOURNE

VAN NOSTRAND REINHOLD COMPANY
25–28 Buckingham Gate, London, SW1E 6LQ

INTERNATIONAL OFFICES
New York   Cincinnati   Toronto   Melbourne

© T. J. FLETCHER 1972

All rights reserved. No part of this publication
may be reproduced, stored in a retrieval system,
or transmitted, in any form or by any means,
electronic, mechanical, photocopying, recording
or otherwise, without the prior permission of
the copyright owner.

Library of Congress Catalog Card No. 72–5920
ISBN   0 442 02411 8 (cloth edition)
         0 442 02410 X (paperback)

First published 1972

Printed in Great Britain by Butler & Tanner Ltd,
Frome and London

To BERYL —
Who believed in this book

Preface

There are already many books on linear algebra, but this one results from an attitude which is far from common. The aim is to approach the theories by way of the problems which the theories can solve, and by way of the situations which they illumine.

The book contains many applications of linear algebra, but it is not a guide to the applications for experts who know the theory already. The approach is the other way about—particular applications in geometry, physics, statistics, and other areas have been chosen because they lead naturally to the key ideas in the theory.

The idea of writing a book on these lines arose from talks given to secondary school teachers. Consequently it is intended for such teachers who are seeking to increase their knowledge of linear algebra and who wish to know where the matrix algebra of recent school syllabuses is leading. The book will also be of value in providing background for students in universities, colleges and in the final years of school who are approaching the subject for the first time and who prefer an approach that is not exclusively theoretical, whether they be students of mathematics or of the physical or social sciences.

Most of the ideas which are the backbone of introductory courses are encountered here. Very many of the essential results are proved, but where rigorous proofs are not possible within the self-imposed limits of the book this is frankly admitted. The first aim is to show why theories are needed and the purposes they serve.

Chapter 1 introduces a variety of situations which, when irrelevant superstructure is discarded, turn out to be linear spaces. Chapter 2 is concerned with fairly familiar co-ordinate geometry, although the balance may seem unusual to some readers. This is because none of this geometry is done for its own sake. The items which are covered have all been chosen because they provide helpful pictures in other contexts in later parts of the book.

Chapters 3 to 5 are a comparatively self-contained unit, and some readers may wish to omit them at first reading. These chapters are an introduction to certain topics in mechanics, especially the analysis of normal modes of vibration. The presentation is intended for readers who already possess some knowledge of mechanics as traditionally taught (in England in the last years of the secondary school), although indications are given of short cuts into the subject for those without this background. In this part of the book the methods of matrix algebra and of differential equations are compared, and gradually brought together, so that the language of operators on linear spaces is developed, and we see how one unified theory can incorporate two areas which hitherto have usually been taught separately. Chapter 5 is the most theoretical, and it is here particularly that rigorous proofs are lacking for some of the main results, although plausible reasons are given. This approach seems justified because the student is meeting abstract linear spaces, a topic which can appear difficult at first acquaintance, and is at the same time being shown the practical considerations from which this branch of mathematics arose.

The usual elementary approach to the eigenvalues of a matrix is by way of determinants and the characteristic equation. This is explained with many examples in Chapter 2; but once beyond the area of elementary exercises quite different methods have to be adopted. Accordingly Chapter 6 is concerned with iterative processes. These processes arise very naturally from the study of certain easily understood problems in probability, but they have a wider importance as algorithms based on them provide practical ways of calculating the eigenvalues and eigenvectors of operators. The essential ideas of linear operators had been understood for a century or more, but their importance increased greatly when the advent of computers made possible the enormous numerical computations which are almost always necessary in the applications. These computations are frequently iterative.

The final chapter covers some further aspects of statistics, especially correlation and the beginnings of multivariate analysis. These show something of the possibilities opened up when once the elements of matrix algebra have been understood. The expert statistician will notice that many difficulties are ignored or glossed over, but the book is not for him. The hope is that the student will be sufficiently interested by the variety of applications in this introduction to proceed afterwards to more rigorous texts.

I would express thanks to many people who have discussed points with me, or made helpful criticism while the book was being written. These include Professor J. V. Armitage, A. W. Bell, I. B. Butterworth, Dr. J. P. Cole, Mrs. B. Fletcher, B. Ford, Professor G. G. Hall, E. H. Larcombe, R. C. Lyness, A. Owen, Dr. G. F. Paechter, and G. F. Peaker.

Finally, I would acknowledge with sincere thanks the kindness of the Shell Centre for Mathematical Education, at Nottingham University, which provided me with opportunities to discuss this work with sympathetic critics and the time to put it down on paper.

Contents

Preface	vii
CHAPTER 1 LINEAR SPACES	1
1.1 Arithmetical progressions	1
1.2 Magic squares	3
1.3 Curve fitting, interpolation and extrapolation	8
1.4 Some other polynomials as a basis	14
1.5 Difference equations and differential equations	20
1.6 Linear dependence in physics and chemistry	23
1.7 Vector spaces over GF(2)	28
1.8 Spaces of infinite dimension	36
1.9 Definitions	38
References	39
CHAPTER 2 SOME ASPECTS OF GEOMETRY	40
2.1 A note on matrix multiplication	40
2.2 Transformations	41
2.3 The solution of simultaneous equations	46
2.4 The systematic use of matrices	53
2.5 Eigenvectors	58
2.6 The inner product and orthogonality	66
2.7 Conics and quadrics	69
2.8 Principal axes and orthogonal transformations	75
2.9 Some further examples on mappings	85
References	96
CHAPTER 3 MECHANICAL VIBRATIONS	97
3.1 Systems of particles	98
3.2 Solution of the general problems	106
3.3 Oscillations in the continuous case	110
3.4 Green's functions	121
3.5 Energy and Rayleigh's principle	127

3.6 A geographical application	134
References	135

Chapter 4 Further Examples — 136

4.1 The revolving chain	136
4.2 Transverse vibrations of beams	139
4.3 The use of difference equations	147
4.4 Circulants	156
4.5 Vibrations in cable nets	158
4.6 Oscillations in electrical filters	159
4.7 Vibrations of drum heads	160
References	163

Chapter 5 Fourier Series and Symmetric Operators — 165

5.1 Projection matrices	165
5.2 Fourier series	167
5.3 Chebyshev polynomials	171
5.4 The Sturm–Liouville equation and symmetric operators	173
5.5 Further properties of the Green's function	186
5.6 Integral operators	189
References	192

Chapter 6 Iterated Powers of Matrices — 194

6.1 A square root process	194
6.2 Beetles	195
6.3 Chemical examples	198
6.4 The educational system	201
6.5 Stochastic matrices	204
6.6 Theory	217
6.7 Root processes continued	224
6.8 Problems involving continuous growth	227
References	230

Chapter 7 Statistical Applications — 232

7.1 Lines of best fit	232
7.2 Multiple regression analysis	237
7.3 Buns, cakes and principal components	241
7.4 Some geographical problems	250
7.5 Miscellaneous examples	252
7.6 An alternative approach to principal components	256
References	260

Solutions and Hints	262
Index	271

I
Linear Spaces

In this chapter we consider a number of situations which are at first sight very different, but investigation shows that they have an underlying similarity of structure. They are all *linear spaces*. Some of these examples serve only as illustrations, and the knowledge that they are linear spaces may give little practical help in handling them. But in other cases the knowledge is useful as it suggests methods of manipulation, and we see that methods which apply in one context are easily transferred to another. The theory of linear spaces receives rigorous discussion in a large number of texts, and so our concern is not to prove theorems but to discuss situations from which the theoretical ideas emerge.

1.1 Arithmetical progressions

It is well known that a sequence of numbers is called an *arithmetical progression* if every pair of successive terms differs by the same amount. Here we are interested in a related, but different property. Experiment with a number of arithmetical progressions will convince us that if two arithmetical progressions are added to one another, term by term, the resulting sequence is again an arithmetical progression.

Thus

$$2,\ 5,\ 8,\ 11,\ 14,\ \ldots$$

added term by term to

$$6,\ 11,\ 16,\ 21,\ 26,\ \ldots$$

gives

$$8,\ 16,\ 24,\ 32,\ 40,\ \ldots\ .$$

When we speak of adding two arithmetical progressions we will always mean addition in this sense.

Arithmetical progressions have another simple property. If any one of them is multiplied, term by term, by a constant the resulting sequence is again an arithmetical progression. Thus

$$2, \ 5, \ 8, \ 11, \ 14, \ \ldots$$

multiplied by 3 gives

$$6, \ 15, \ 24, \ 33, \ 42, \ \ldots \ .$$

This operation is called *scalar multiplication*.

To summarize, we may say that the set of arithmetical progressions is *closed* with respect to the operations of: (i) addition term by term, and (ii) scalar multiplication. Many other mathematical systems behave in this way, and any system which has these two properties we call a *vector space* or a *linear space*. The elements of such a space we will call *vectors*, so that in the particular space in this example each vector is an arithmetical progression. By using these two operations various things can be done; in particular, any arithmetical progression may be expressed in terms of two special progressions,

$$\mathbf{x} = 1, \ 1, \ 1, \ 1, \ \ldots$$

and

$$\mathbf{y} = 0, \ 1, \ 2, \ 3, \ \ldots \ .$$

(Note that $\mathbf{x}$ and $\mathbf{y}$ are used as abbreviations for the complete sequences.)

Thus the sequence

$$6, \ 11, \ 16, \ 21, \ 26, \ \ldots$$

is expressible as $6\mathbf{x} + 5\mathbf{y}$; that is it is expressible linearly in terms of $\mathbf{x}$ and $\mathbf{y}$.

The two vectors $\mathbf{x}$ and $\mathbf{y}$ may be called a *basis* for the space.

It is quite possible to select a different pair of arithmetical progressions as a basis and to express others in terms of them (try this), but the sequences $\mathbf{x}$ and $\mathbf{y}$ are an especially simple pair to use. Since any other arithmetical progression may be expressed in terms of a special *two*, we say that the space of arithmetical progressions is *two-dimensional*.

Exercise 1.1

Consider any three arithmetical progressions; show that it is always possible to express one linearly in terms of the other two.

Exercise 1.2

Represent the arithmetical progression $p\mathbf{x} + q\mathbf{y}$ by the point (p, q) in a Cartesian plane. Which points represent the arithmetical progressions which are scalar multiples of the one represented by (p, q)? How is addition of arithmetical progressions represented in the Cartesian plane?

1.2 Magic squares

Most people recognize
$$\begin{bmatrix} 4 & 9 & 2 \\ 3 & 5 & 7 \\ 8 & 1 & 6 \end{bmatrix}$$
as a magic square, and the properties of magic squares are well known:

(i) every row has the same total;
(ii) every column has the same total again;
(iii) the two diagonals have the same total again.

Some magic squares have another property;

(iv) the components are consecutive integers.

In the present context we cannot handle property (iv), and we will *not* take this as part of our definition of 'magic square'. Sometimes it is interesting to consider squares of numbers with properties (i) and (ii) only. Such squares will be called *semi-magic*. Squares of numbers with properties (i), (ii) and (iii) will be called *magic*.

It is easy to see that if two magic squares are added element by element the resulting square is magic. If a magic square is multiplied element by element by a constant, the result is magic. Hence in this context also we may speak of addition of squares and scalar multiplication of squares, and regard the set of all magic squares of some particular size as a space. If we are considering 3×3 magic squares, what is the dimension of this space?

This is not as simple as the previous example, and most people find it easier to start by finding the dimension of the space of semi-magic squares.

Can we find some especially simple semi-magic squares with a view to using them as base vectors?

$$\mathbf{a} = \begin{bmatrix} 1 & 1 & 1 \\ 1 & 1 & 1 \\ 1 & 1 & 1 \end{bmatrix}, \quad \mathbf{b} = \begin{bmatrix} 1 & 0 & 0 \\ 0 & 1 & 0 \\ 0 & 0 & 1 \end{bmatrix}, \quad \mathbf{c} = \begin{bmatrix} 0 & 0 & 1 \\ 0 & 1 & 0 \\ 1 & 0 & 0 \end{bmatrix}$$

are easy to find, and the first is magic as well as semi-magic. Further simple ones are

$$\mathbf{d} = \begin{bmatrix} 1 & 0 & 0 \\ 0 & 0 & 1 \\ 0 & 1 & 0 \end{bmatrix}, \quad \mathbf{e} = \begin{bmatrix} 0 & 0 & 1 \\ 1 & 0 & 0 \\ 0 & 1 & 0 \end{bmatrix},$$

$$\mathbf{f} = \begin{bmatrix} 0 & 1 & 0 \\ 1 & 0 & 0 \\ 0 & 0 & 1 \end{bmatrix}, \quad \mathbf{g} = \begin{bmatrix} 0 & 1 & 0 \\ 0 & 0 & 1 \\ 1 & 0 & 0 \end{bmatrix}.$$

How long is it before we start finding that some are not 'really new' because they can be expressed linearly in terms of squares which we have already? Inspection shows that

$$\mathbf{a} = \mathbf{d} + \mathbf{f} + \mathbf{c} = \mathbf{e} + \mathbf{g} + \mathbf{b}.$$

This shows that **a** is redundant and that any one of the remaining six can be expressed in terms of the other five. Two questions now arise. Can we express *any* other semi-magic square in terms of the ones we have found so far? Are there any further relations which lead us to say that fewer than five can be taken as basic, because one of these can be expressed in terms of the other four?

First we will show that any semi-magic square can be expressed in terms of **b, d, e, f** and **g**. The square at the beginning of this section will illustrate a procedure which can always be employed. This square has 5 in the centre. The only one of **b, d, e, f** and **g** which has a non-zero element in the centre is **b**, which has 1. So we must have 5**b**. **d** is the only other basic square with an element different from zero in the top left-hand corner; so to get the 4 in the top left-hand corner of the square we are trying to form we must subtract **d**. Arguing from the other corners we see successively that we must have 2**e**, 8**g** and **f**. Thus the square we started with is equal to $5\mathbf{b} - \mathbf{d} + 2\mathbf{e} + \mathbf{f} + 8\mathbf{g}$.

It may be verified that any semi-magic square may be expressed in terms of **b, d, e, f** and **g** in this way; but some details are left for the reader to fill in. For example, the four elements at the centres of the sides have not come into the calculation. Are they 'forced' if the other five elements are determined?

Now can any of **b, d, e, f** and **g** be expressed linearly in terms of the rest? **b** clearly cannot be expressed in terms of the others because they all have zero at the centre and **b** has unity. Similar arguments on corner elements show that neither **e** nor **g** can be expressed in terms of the others. A slight variation of the argument can be applied in the case of **d** and **f**.

We therefore conclude that any semi-magic square can be expressed in terms of five which we have taken as basic, and that it will not do to take only four out of the five. This being so, we may say that the space of semi-magic squares is of dimension 5.

But there is a little more to investigate. Someone else might make a completely different start on the problem. He might consider, for example,

$$\mathbf{u} = \begin{bmatrix} 1 & -1 & 0 \\ -1 & 1 & 0 \\ 0 & 0 & 0 \end{bmatrix}, \quad \mathbf{v} = \begin{bmatrix} 0 & 1 & -1 \\ 0 & -1 & 1 \\ 0 & 0 & 0 \end{bmatrix},$$

$$\mathbf{w} = \begin{bmatrix} 0 & 0 & 0 \\ 1 & -1 & 0 \\ -1 & 1 & 0 \end{bmatrix} \quad \text{and} \quad \mathbf{x} = \begin{bmatrix} 0 & 0 & 0 \\ 0 & 1 & -1 \\ 0 & -1 & 1 \end{bmatrix}.$$

These squares are all semi-magic and from them others may be derived by linear operations. Do we need to include others which are rather similar? For example, consider

$$\mathbf{y} = \begin{bmatrix} 1 & 0 & -1 \\ 0 & 0 & 0 \\ -1 & 0 & 1 \end{bmatrix}.$$

It is easy to see that $\mathbf{y} = \mathbf{u} + \mathbf{v} + \mathbf{w} + \mathbf{x}$, so $\mathbf{y}$ is not an independent semi-magic square if we have already selected $\mathbf{u}$, $\mathbf{v}$, $\mathbf{w}$ and $\mathbf{x}$.

Exercise 1.3

How can **a**, **u**, **v**, **w** and **x** be expressed linearly in terms of **b**, **d**, **e**, **f** and **g**?

Can **b**, **d**, **e**, **f** and **g** be expressed linearly in terms of **a**, **u**, **v**, **w** and **x**?

Can any of **a**, **u**, **v**, **w** and **x** be expressed linearly in terms of the others?

By experimenting in this way one is led to the essential ideas of dimension. If two different investigators set about the task of selecting certain squares to use as a basis, and if they do the natural thing and each leaves out of his set of basic squares any which are expressible in terms of those which he has decided are basic already, then they end up eventually each with the same number of basic squares, although the squares which they select may be different. This number is called the *dimension* of the system. It is, of course, necessary when developing the theory to be able to provide a logical proof of this fact about dimension. Rigorous proofs of this kind of result are to be found in standard texts such as those by Birkhoff and MacLane [1] and Mirsky [10].

The value of an established nomenclature should now be clear. In the first example sequences were added, in the second example semi-magic squares. Later we will add polynomials and, afterwards, still more complicated mathematical functions. When different things are to be processed in the same way it is convenient to adopt a common name. This brings out the importance of the terminology adopted earlier. Any mathematical objects which may be (i) added together and (ii) multiplied by scalars, may be called vectors. The set of vectors together with the two operations is called a vector space or linear space.

'What *is* a vector?' is not really a proper question. The proper question is 'How does a vector behave?' Mathematical definitions nowadays answer the second type of question far more often than they answer the first. A more formal definition of a linear space is given at the end of the chapter.

The reader may now investigate the space of *magic* squares in a similar way. Various squares may be taken as 'basic'. Three squares which are a possible choice as a basis are

$$\begin{bmatrix} 1 & 1 & 1 \\ 1 & 1 & 1 \\ 1 & 1 & 1 \end{bmatrix} \quad \begin{bmatrix} 0 & 1 & -1 \\ -1 & 0 & 1 \\ 1 & -1 & 0 \end{bmatrix} \text{ and } \begin{bmatrix} -1 & 1 & 0 \\ 1 & 0 & -1 \\ 0 & -1 & 1 \end{bmatrix}.$$

Exercise 1.4

Show that the space of magic squares is of dimension three.

Another piece of terminology arises very naturally from the problem. The space of magic squares is a sub-space of the space of semi-magic squares. We say that one space is a sub-space of another if the first is contained in the second (obviously enough), but it is important also that it should be a *space* in its own right; that is to say that it must be closed with respect to addition and scalar multiplication—these two operations must never carry us outside the system. The idea of dimension in a vector space and the idea that there are various possible base vectors in a vector space are very important, and may be illustrated further by considering 4×4 magic squares. A note by Botsch in a German journal [2] shows how 4×4 magic squares can be constructed by taking linear sums of the eight basic squares:

$$\begin{bmatrix} 1 & 0 & 0 & 0 \\ 0 & 0 & 0 & 1 \\ 0 & 1 & 0 & 0 \\ 0 & 0 & 1 & 0 \end{bmatrix}, \begin{bmatrix} 0 & 1 & 0 & 0 \\ 0 & 0 & 1 & 0 \\ 1 & 0 & 0 & 0 \\ 0 & 0 & 0 & 1 \end{bmatrix}, \begin{bmatrix} 0 & 0 & 0 & 1 \\ 0 & 1 & 0 & 0 \\ 1 & 0 & 0 & 0 \\ 0 & 0 & 1 & 0 \end{bmatrix}, \begin{bmatrix} 0 & 1 & 0 & 0 \\ 0 & 0 & 0 & 1 \\ 0 & 0 & 1 & 0 \\ 1 & 0 & 0 & 0 \end{bmatrix},$$

$$\begin{bmatrix} 0 & 0 & 0 & 1 \\ 1 & 0 & 0 & 0 \\ 0 & 0 & 1 & 0 \\ 0 & 1 & 0 & 0 \end{bmatrix}, \begin{bmatrix} 0 & 0 & 1 & 0 \\ 0 & 1 & 0 & 0 \\ 0 & 0 & 0 & 1 \\ 1 & 0 & 0 & 0 \end{bmatrix}, \begin{bmatrix} 1 & 0 & 0 & 0 \\ 0 & 0 & 1 & 0 \\ 0 & 0 & 0 & 1 \\ 0 & 1 & 0 & 0 \end{bmatrix}, \begin{bmatrix} 0 & 0 & 1 & 0 \\ 1 & 0 & 0 & 0 \\ 0 & 1 & 0 & 0 \\ 0 & 0 & 0 & 1 \end{bmatrix}.$$

These squares are all magic. They are, incidentally, interrelated by the symmetries of a square; given any one, the others may be generated by carrying out the symmetry operations of the group of the square on it. The eight are not independent as it is easy to see that the sum of the left-hand four is equal to the sum of the right-hand four. A little further investigation will show that there are no further linear dependence relations between them, and so they span a space of dimension seven. Botsch pointed out that a famous magic square:

$$\begin{bmatrix} 16 & 3 & 2 & 13 \\ 5 & 10 & 11 & 8 \\ 9 & 6 & 7 & 12 \\ 4 & 15 & 14 & 1 \end{bmatrix}$$

occurs in the woodcut by Dürer entitled 'Melancholy', and that this magic square can be expressed as a linear sum of any seven of the eight given above.

Exercise 1.5
Prove this.

But are there any 4×4 magic squares which cannot be expressed in this way? Dr. G. Barratt sent to me the results of an investigation which included eight 4×4 magic squares, the last four of Botsch's together with

$$\begin{bmatrix} \frac{1}{2} & 1 & 0 & -\frac{1}{2} \\ 0 & 0 & 0 & 0 \\ 0 & 0 & 0 & 0 \\ \frac{1}{2} & -1 & 0 & \frac{1}{2} \end{bmatrix}, \begin{bmatrix} \frac{1}{2} & 0 & 0 & -\frac{1}{2} \\ -1 & 0 & 0 & 1 \\ 0 & 0 & 0 & 0 \\ \frac{1}{2} & 0 & 0 & -\frac{1}{2} \end{bmatrix},$$

$$\begin{bmatrix} \frac{1}{2} & 0 & -1 & \frac{1}{2} \\ 0 & 0 & 0 & 0 \\ 0 & 0 & 0 & 0 \\ -\frac{1}{2} & 0 & 1 & -\frac{1}{2} \end{bmatrix}, \begin{bmatrix} -\frac{1}{2} & 0 & 0 & \frac{1}{2} \\ 0 & 0 & 0 & 0 \\ 1 & 0 & 0 & -1 \\ -\frac{1}{2} & 0 & 0 & \frac{1}{2} \end{bmatrix},$$

These eight magic squares are all independent. Can you see a quick way of proving this without elaborate calculation? The fact that Barratt has produced a space of magic squares of dimension eight when Botsch produced a space of dimension seven suggests that Barratt's method constructs some which cannot be constructed by Botsch's method. We say *suggests* here because we have not proved any general results about the theory of dimension. It is the purpose of general theories to justify conclusions of this nature.

Exercise 1.6
Show that at least one of Barratt's squares is not constructible as a linear sum of Botsch's squares.
Show that all of Botsch's squares are constructible as linear sums of Barratt's. (This shows that Botsch's space is a sub-space of Barratt's.)
Show that Dürer's square is constructible as a linear sum of Barratt's.
Show that any 4×4 magic square is constructible as a linear sum of Barratt's.

The reader is invited to attempt these questions arguing from first principles, but the results can all be deduced from a basic general theorem, to which we have already referred:

Any two sets of independent vectors which are a basis for a space contain the same number of vectors.

Exercise 1.7
Can you construct a magic pentagram? That is to say, can you put numbers in the circles shown in Fig. 1.1 in such a way that the total in

every line is the same? What is the dimension of the space of magic pentagrams?

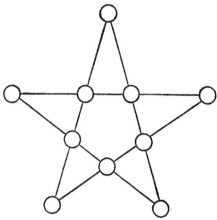

FIG. 1.1

Exercise 1.8
What can you say about magicubes?

1.3 Curve fitting, interpolation and extrapolation

In elementary mathematics we frequently need to calculate with quadratic functions. At first sight a quadratic function of x is clearly composed of a term in x^2, a term in x and a constant term. (We include the cases when coefficients are zero.) When thinking this way we are effectively using three special functions, x^2, x and 1, as a basis for expressing all quadratic functions. In some circumstances other quadratic functions turn out to be more appropriate as a basis; it depends on the job in hand.

Consider the problem of fitting a quadratic curve through three points. Suppose that we are given the values of $f(x)$ when x has the values 0, 1 and 2 and we wish to determine the quadratic function of x which takes these values at these points. If the job is to be done once only then it is reasonable to put $f(x) = ax^2 + bx + c$ and to determine a, b and c by substitution and by solving the simultaneous equations. However, if this has to be done over and over again with different sets of values at the points $x = 0, 1, 2$, then there is a better way. Let us form three auxiliary quadratic functions,

$f_1(x)$, which has the value 1 when $x = 0$ and is zero at $x = 1$ and $x = 2$, and
$f_2(x)$, which has the value 1 when $x = 1$ and is zero at $x = 0$ and $x = 2$, and
$f_3(x)$, which has the value 1 when $x = 2$ and is zero at $x = 0$ and $x = 1$.

1] CURVE FITTING, INTERPOLATION AND EXTRAPOLATION

These are easily calculated:
$$f_1(x) = \tfrac{1}{2}(x-1)(x-2) = \tfrac{1}{2}x^2 - \tfrac{3}{2}x + 1,$$
$$f_2(x) = -x(x-2) = -x^2 + 2x,$$
$$f_3(x) = \tfrac{1}{2}x(x-1) = \tfrac{1}{2}x^2 - \tfrac{1}{2}x.$$

The function $f(x)$ which takes values a, b and c at $x = 0$, 1 and 2 respectively, is then
$$f(x) = af_1(x) + bf_2(x) + cf_3(x).$$

We may say that we are working in a three-dimensional vector space, with base vectors $f_1(x)$, $f_2(x)$ and $f_3(x)$. This system is a vector space because the elements being used can be handled in the two essential ways, they can be added appropriately and they can be multiplied by scalar constants. This system can also be called a *function space* because the elements are functions. Other function spaces will be met with later.

EXAMPLE 1.1

The polynomials $f_1(x)$, $f_2(x)$ and $f_3(x)$ are called *Lagrange interpolation polynomials*. What are the appropriate Lagrange polynomials when the values a, b, c of the function are to be assigned at $x = -1$, 0, +1? The required functions are
$$f_1(x) = \tfrac{1}{2}x(x-1) = \tfrac{1}{2}x^2 - \tfrac{1}{2}x,$$
$$f_2(x) = -(x-1)(x+1) = -x^2 + 1,$$
$$f_3(x) = \tfrac{1}{2}x(x+1) = \tfrac{1}{2}x^2 + \tfrac{1}{2}x.$$

EXAMPLE 1.2

If we know that the altitudes of a rocket at times $t = 1, 2, 3$, are a, b and c respectively, what are reasonable estimates for its altitudes at $t = 4$, 5 and 10? The data given will suffice to fit a quadratic approximation. Using the appropriate Lagrange interpolation polynomials we may write, if the height is h,
$$h = \tfrac{1}{2}a(t-2)(t-3) - b(t-1)(t-3) + \tfrac{1}{2}c(t-1)(t-2).$$

When $t = 4$, $h = a - 3b + 3c$.
When $t = 5$, $h = 3a - 8b + 6c$.
When $t = 10$, $h = 28a - 63b + 36c$.

Here the *interpolation* polynomials are being used as *extrapolation* polynomials; the example is only an illustration and it may be very crude as rocket technology. In such applications of mathematics there must always be a skilled appraisal of the approximations involved and of the relevance of the mathematics to the real problem.

EXAMPLE 1.3

The velocity of a bullet at $t = 0$ is v, and its distances s at times $t = 1$ and $t = 2$ are a and b respectively. What is a reasonable estimate for its distance at $t = 3$? On the evidence available what can be said about its retardation? Since it is the *velocity* which is given at $t = 0$ the technique must be modified. Suppose

$$s = pt^2 + qt + r,$$

then
$$s' = 2pt + q,$$

the prime denoting differentiation with respect to t. Hence

$$q = v,$$
and
$$a = p + q + r,$$
and
$$b = 4p + 2q + r.$$

Solving
$$p = \tfrac{1}{3}(b - a - v),$$
and
$$r = \tfrac{1}{3}(4a - b - 2v).$$

So in general

$$s = \tfrac{1}{3}(b - a - v)t^2 + vt + \tfrac{1}{3}(4a - b - 2v),$$
$$= (-\tfrac{1}{3}t^2 + \tfrac{4}{3})a + (\tfrac{1}{3}t^2 - \tfrac{1}{3})b + (-\tfrac{1}{3}t^2 + t - \tfrac{2}{3})v.$$

When $t = 3$,
$$s = -\tfrac{5}{3}a + \tfrac{8}{3}b - \tfrac{2}{3}v.$$

The obviously reasonable way to estimate the retardation from the data is to use the second derivative of s, which is equal to

$$2p = \tfrac{2}{3}(b - a - v).$$

Exercise 1.9

Consider similar problems in which the times are known for various distances, as is the case with certain electronic timing devices. Could similar methods be used?

EXAMPLE 1.4

Consider now the problems which arise when a moving object is filmed and we wish to estimate its position at times between the instants when successive pictures on the film are taken, and also its velocities and its accelerations. For convenience we may imagine the object moving past a fence with regularly spaced posts, or some similar situation in which measurement of the distances is easy. Similar problems arise in the school science laboratory when tape is pulled

1] CURVE FITTING, INTERPOLATION AND EXTRAPOLATION 11

through a 'ticker' device which makes marks on it at regular intervals of time.

First use three successive pictures on the film to estimate the velocity and the acceleration at the instant when the middle picture is taken. It is convenient to use the polynomials in Example 1.1, which involves taking the origin of time when the middle picture is taken. We have

$$s = \tfrac{1}{2}a(t^2 - t) - b(t^2 - 1) + \tfrac{1}{2}c(t^2 + t),$$

where a, b and c are the displacements measured on the film or on the ticker-tape at times $t = -1$, 0 and 1, respectively.

The velocity v is given by

$$v = \tfrac{1}{2}a(2t - 1) - 2bt + \tfrac{1}{2}c(2t + 1),$$

and when $t = 0$

$$v = \tfrac{1}{2}(c - a).$$

The acceleration f is given by

$$f = a - 2b + c.$$

This is independent of t because the assumption that a quadratic curve will fit the data is equivalent to the assumption that the acceleration is constant.

As a further application of the same ideas we may find the area under the curve

$$y = \tfrac{1}{2}a(x^2 - x) - b(x^2 - 1) + \tfrac{1}{2}c(x^2 + x),$$

between $x = -1$ and $x = +1$.

Integrating, we find that

$$\text{Area} = [\tfrac{1}{2}a(\tfrac{1}{3}x^3 - \tfrac{1}{2}x^2) - b(\tfrac{1}{3}x^3 - x) + \tfrac{1}{2}c(\tfrac{1}{3}x^3 + \tfrac{1}{2}x^2)]_{-1}^{+1},$$

$$= \tfrac{1}{3}(a + 4b + c).$$

This formula is the foundation for the method of estimating areas under curves by Simpson's rule.

EXAMPLE 1.5

It is interesting to move a little beyond the restrictions of the previous example and to consider what use can be made of the information on four successive frames of film, or from the positions of four successive dots on timing ticker-tape. Assuming that the distances s, at $t = -\tfrac{3}{2}$, $-\tfrac{1}{2}$, $+\tfrac{1}{2}$, $+\tfrac{3}{2}$, are a, b, c and d respectively, we will estimate the distance, velocity and acceleration at $t = 0$.

As a first step the four appropriate interpolation polynomials must be calculated. These are

$$s_1(t) = -\tfrac{1}{6}(t+\tfrac{1}{2})(t-\tfrac{1}{2})(t-\tfrac{3}{2}) = -\tfrac{1}{48}(8t^3 - 12t^2 - 2t + 3),$$
$$s_2(t) = \tfrac{1}{2}(t+\tfrac{3}{2})(t-\tfrac{1}{2})(t-\tfrac{3}{2}) = \tfrac{1}{16}(8t^3 - 4t^2 - 18t + 9),$$
$$s_3(t) = -\tfrac{1}{2}(t+\tfrac{3}{2})(t+\tfrac{1}{2})(t-\tfrac{3}{2}) = -\tfrac{1}{16}(8t^3 + 4t^2 - 18t - 9),$$
$$s_4(t) = \tfrac{1}{6}(t+\tfrac{3}{2})(t+\tfrac{1}{2})(t-\tfrac{1}{2}) = \tfrac{1}{48}(8t^3 + 12t^2 - 2t - 3).$$

In the above working the polynomials are first written in factorized form, as their roots are known, and then the coefficient is chosen in such a way as to make the value unity at the fourth given point.

Then
$$s = as_1(t) + bs_2(t) + cs_3(t) + ds_4(t).$$

At $t = 0$,
$$s = \tfrac{1}{16}(9b + 9c - a - d).$$

The velocity, v, is given by
$$v = -\frac{a}{24}(12t^2 - 12t - 1) + \frac{b}{8}(12t^2 - 4t - 9)$$
$$-\frac{c}{8}(12t^2 + 4t - 9) + \frac{d}{24}(12t^2 + 12t - 1).$$

When $t = 0$
$$v = \frac{1}{24}(a - d + 27c - 27b).$$

This may be rearranged as
$$v = \frac{9}{8}(c - b) - \frac{1}{8}\left(\frac{d - a}{3}\right),$$

showing how this estimate of the velocity may be regarded as a weighted mean obtained from the estimates of the velocity given by the middle two readings and by the outer two readings. The first estimate is obviously the closer of the two, but it is interesting to see that the estimate from the outer two has to be used with a negative weight. The acceleration is given by

$$f = -\frac{a}{6}(6t - 3) + \frac{b}{2}(6t - 1) - \frac{c}{2}(6t + 1) + \frac{d}{6}(6t + 3).$$

When $t = 0$
$$f = \tfrac{1}{2}(a - b - c + d).$$

This may be rearranged as
$$f = \tfrac{1}{2}[(a + c - 2b) + (b + d - 2c)];$$

1] CURVE FITTING, INTERPOLATION AND EXTRAPOLATION 13

that is to say the acceleration at the mid-point of the four observations is given as the mean of the accelerations obtained using the first three points only and the last three only.

The polynomials may also be used to evaluate the area under the curve

$$y = as_1(x) + bs_2(x) + cs_3(x) + ds_4(x).$$

Between $x = -\frac{3}{2}$ and $x = +\frac{3}{2}$ the result is

$$\text{Area} = \tfrac{3}{8}(a + 3b + 3c + d).$$

This may be used to develop a formula giving a higher order approximation to the area under curves than that provided by Simpson's rule.

The above calculations involve expressing cubic functions in terms of four base functions. The 'space of cubics' is of dimension four.

EXAMPLE 1.6

This example introduces in a simplified form ideas which can be applied to the production of graphical outputs from computers. As the result of experimental work or numerical calculations the values of a function may be known at a number of points, and we may wish to join these points by a continuous curve such as is shown in Fig. 1.2. Many criteria of fit may be chosen. One among many possible methods is to use the values of the function at three consecutive points, call them $x = 0$, $x = 1$ and $x = 2$, and to match the gradient of the curve which

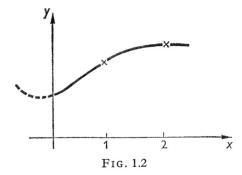

FIG. 1.2

we will assume has been drawn up to $x = 0$ so far. This gives an interpolating cubic which can be used to extend the curve to $x = 2$.

The value of the gradient at $x = 2$ must then be calculated and this is used in defining the fresh cubic curve which is used to draw the curve over the interval (2, 4). Of course, it is convenient to re-label the co-ordinates at each stage and so always work effectively over the interval (0, 2).

Let the given data be

$$f(0) = a, \quad f(1) = b, \quad f(2) = c, \quad \text{and} \quad f'(0) = d.$$

It is desirable to define the base cubics by

$$f_1(0) = 1, \quad f_1(1) = 0, \quad f_1(2) = 0, \quad \text{and} \quad f_1'(0) = 0,$$
$$f_2(0) = 0, \quad f_2(1) = 1, \quad f_2(2) = 0, \quad \text{and} \quad f_2'(0) = 0,$$
$$f_3(0) = 0, \quad f_3(1) = 0, \quad f_3(2) = 1, \quad \text{and} \quad f_3'(0) = 0,$$
$$\text{and} \quad f_4(0) = 0, \quad f_4(1) = 0, \quad f_4(2) = 0, \quad \text{and} \quad f_4'(0) = 1.$$

These functions turn out to be

$$f_1(x) = \tfrac{1}{4}(3x^3 - 7x^2 + 4),$$
$$f_2(x) = -x^3 + 2x^2,$$
$$f_3(x) = \tfrac{1}{4}(x^3 - x^2),$$
$$f_4(x) = \tfrac{1}{2}(x^3 - 3x^2 + 2x).$$

It is worth sketching these curves and considering their plausibility in physical terms, because to a first approximation (i.e. for small departures from the x-axis) they are the shapes taken up by flexible metal strips which are constrained to satisfy the imposed conditions.

The required interpolation curve is

$$f(x) = af_1(x) + bf_2(x) + cf_3(x) + df_4(x).$$

$$f'(x) = \frac{a}{4}(9x^2 - 14x) + b(-3x^2 + 4x) + \frac{c}{4}(3x^2 - 2x) + \frac{d}{2}(3x^2 - 6x + 2),$$

and $f'(2) = 2a - 4b + 2c + d$,

and this is part of the data for the construction of the next length of curve.

It is interesting to consider the meaning of the various components of the expression in terms of the flexible strip model. Consider especially the signs of the terms and check their plausibility.

1.4 Some other polynomials as a basis

The main point of the last section was to show that when dealing with polynomials it may be convenient to take certain polynomials as base polynomials and express others in terms of them. Which particular polynomials are best regarded as basic depends entirely upon circumstances. In this section a different starting point leads to the adoption of a new set of polynomials as base vectors for the vector space of polynomials. The topic is suggested by Sawyer [14].

Difference tables are usually studied early in any course on numerical methods, and certainly find a place in a number of school

SOME OTHER POLYNOMIALS AS A BASIS

texts. An example will explain the idea. We may list the values which $y = x^2 + 4x + 3$ takes for $x = 0, 1, 2, \ldots$. The values are

$$3 \quad 8 \quad 15 \quad 24 \quad 35 \quad 48, \text{ etc.}$$

In a line below we record the differences between the terms listed before.

$$3 \quad 8 \quad 15 \quad 24 \quad 35 \quad 48 \ldots$$
$$5 \quad 7 \quad 9 \quad 11 \quad 13 \ldots$$

We may carry on like this indefinitely, forming differences of differences, etc., until we obtain

$$3 \quad 8 \quad 15 \quad 24 \quad 35 \quad 48 \ldots$$
$$5 \quad 7 \quad 9 \quad 11 \quad 13 \ldots$$
$$2 \quad 2 \quad 2 \quad 2 \ldots$$
$$0 \quad 0 \quad 0 \ldots$$

Once a line of zeros has appeared the task is complete, because any further lines which may be constructed will consist of zeros also.

Consider another example, tabulating the values of $y = x^3 + 1$ for $x = 0, 1, 2, \ldots$ and the successive differences.

$$1 \quad 2 \quad 9 \quad 28 \quad 65 \quad 126 \ldots$$
$$1 \quad 7 \quad 19 \quad 37 \quad 61 \ldots$$
$$6 \quad 12 \quad 18 \quad 24 \ldots$$
$$6 \quad 6 \quad 6 \ldots$$
$$0 \quad 0 \ldots$$

Again a row of zeros is produced eventually. It will be found that starting with any polynomial of degree n, after $n + 1$ rows all subsequent rows are zero.

Exercise 1.10
  Prove this.

The use of tables like these in numerical computation is explained in textbooks, but here we will restrict attention merely to one or two aspects. From a numerical table of this kind one might wish to calculate the polynomial on which the table is based. This might arise in an experimental situation where we have only the numerical data and we wish to find the polynomial which fits.

Given the numerical table, how may the corresponding polynomial be found? One quite convenient method arises from the following considerations.

Tables may be added term by term, as vectors. Thus, adding the two tables previously considered we get

$$\begin{array}{cccccc} 4 & 10 & 24 & 52 & 100 & 174 \quad \ldots \\ 6 & 14 & 28 & 48 & 74 & \ldots \\ 8 & 14 & 20 & 26 & \ldots \\ 6 & 6 & 6 & \ldots \\ 0 & 0 & \ldots \end{array}$$

and this is simply the table for the polynomial $y = x^3 + x^2 + 4x + 4$. The set of tables of this kind is a vector space; can we find a convenient basis?

Notice next that since each table contains many sets of three numbers, arranged in triangles:

$$\begin{array}{cc} x & y \\ z & \end{array} \quad \text{with } z = y - x,$$

we know also that $y = x + z$, and the table may be built up from the left-hand edge by successive additions, as well as from the top row by successive subtractions. Many algebraic relations arise from this.

Knowing that the numbers down the left-hand edge determine the rest of the table, and following the ideas which were useful before we see that it might help to consider tables with especially simple left-hand edges, i.e. the tables commencing

$$\begin{array}{cccc} 1 \ldots & 0 \ldots & 0 \ldots & 0 \ldots \\ 0 \ldots & 1 \ldots & 0 \ldots & 0 \ldots \\ 0 \ldots & 0 \ldots & 1 \ldots & 0 \ldots \quad \text{etc.} \\ 0 & 0 & 0 & 1 \end{array}$$

What polynomials correspond to these? The general rule is easily seen if we consider the fourth of these as a typical case. Filling in successive entries in this table we see that it begins

$$\begin{array}{cccccc} 0 & 0 & 0 & 1 & 4 & 10 \quad \ldots \\ 0 & 0 & 1 & 3 & 6 & \ldots \\ 0 & 1 & 2 & 3 & \ldots \\ 1 & 1 & 1 & \ldots \\ 0 & 0 & \ldots \end{array}$$

The numbers appearing will be familiar as the entries in Pascal's triangle. Using the results of Exercise 1.10 we know that this is the difference table of a polynomial $y = f(x)$, of degree three. Furthermore, we see at once that

$$f(0) = f(1) = f(2) = 0 \quad \text{and} \quad f(3) = 1.$$

SOME OTHER POLYNOMIALS AS A BASIS

The polynomial is therefore
$$y = x(x-1)(x-2)/6.$$

Writing this as
$$f_3(x) = x(x-1)(x-2)/3!,$$
it is a simple matter to verify that the polynomials corresponding to the successive basic tables are

$$f_0(x) = 1,$$
$$f_1(x) = x/1!,$$
$$f_2(x) = x(x-1)/2!,$$
$$f_3(x) = x(x-1)(x-2)/3!,$$
$$\cdots \cdots \cdots$$
$$f_r(x) = x(x-1)(x-2)\ldots(x-r+1)/r!.$$

These polynomials are especially helpful when it comes to identifying the polynomial which is related to a particular difference table. Thus referring back to the table with first row (1, 2, 9, 28, 65, 126, …), it had at the left-hand edge (1, 1, 6, 6, 0, …). Hence the related polynomial is

$$f(x) = f_0(x) + f_1(x) + 6f_2(x) + 6f_3(x),$$
$$= 1 + x + 3x(x-1) + x(x-1)(x-2),$$
$$= x^3 + 1,$$

which we know to be correct.

When x is an integer greater than r the functions $f_r(x)$ are familiar as binomial coefficients, when they are more usually written in terms of the variables n and r with the notation ^nC_r.

Algebra texts contain large numbers of relevant examples, for some of which the present vector space point of view is helpful. The functions we have just introduced are also useful in summing series, because summing terms in the top row of a difference table is merely a matter of inserting another row of the table above the existing top row, putting a zero at the left-hand end. Thus, consider the table for $y = x^2$, which is

```
0   1   4   9   16   25  ...
  1   3   5   7    9    ...
    2   2   2   2      ...
      0   0   0        ...
```

If another row is inserted above, starting with a zero, the law of formation extends the table to

$$
\begin{array}{ccccccc}
0 & 0 & 1 & 5 & 14 & 30 & 55 \ \ldots \\
0 & 1 & 4 & 9 & 16 & 25 & \ldots \\
1 & 3 & 5 & 7 & 9 & \ldots & \\
2 & 2 & 2 & 2 & \ldots & &
\end{array}
$$

The terms appearing in the first row, after the first term, are the partial sums of the series

$$0 + 1 + 4 + 9 + 16 \ldots.$$

We know that this is the table for the function

$$f(x) = f_2(x) + 2f_3(x) = \tfrac{1}{2}x(x-1) + \tfrac{1}{3}x(x-1)(x-2)$$
$$= \tfrac{1}{6}x(x-1)(2x-1).$$

Noting carefully the labelling conventions, this formula is easily seen to be the well-known one for the summation of squares, in the form

$$\sum_{r=0}^{n-1} r^2 = \tfrac{1}{6}(n-1)n(2n-1).$$

The more familiar form of this is

$$\sum_{r=1}^{n} r^2 = \tfrac{1}{6}n(n+1)(2n+1).$$

Exercise 1.11

Use these methods to derive the formulae

$$\sum_{r=1}^{n} r = \tfrac{1}{2}n(n+1),$$

$$\sum_{r=1}^{n} r^3 = [\tfrac{1}{2}n(n+1)]^2.$$

Exercise 1.12

The successive values of $f_3(x)$ are

0, 0, 0, 1, 4, 10, 20, 35 ... .

Verify that, labelling from a suitable point, these are the tetrahedral numbers which arise by summing the triangular numbers, these being the successive values of $f_2(x)$, again labelling from a suitable point.

More generally, show that

$$f_k(0) + f_k(1) + \ldots + f_k(n) = f_{k+1}(n+1) \qquad (1.1)$$

This is a useful summation formula because with its aid sums of the form $\sum_{x=0}^{x=n} f(x)$, where $f(x)$ is a polynomial, can be calculated. $f(x)$ has first to be expressed as a linear sum of polynomials of the form $f_k(x)$, and then Eq. (1.1) may be applied to each $f_k(x)$. The point is that a sum of the form $\sum_{x=0}^{x=n} f_k(x)$ is easily found, as Eq. (1.1) always applies, whereas sums of the form $\sum_{x=0}^{x=n} x^k$ are awkward to calculate and the formulae become increasingly complicated with increasing values of k. Therefore, for this purpose, the calculation of partial sums, it is much more convenient to think of a polynomial as a linear sum of the form

$$f(x) = \sum_k a_k f_k(x)$$

than as a linear sum of the form

$$f(x) = \sum_k b_k x^k.$$

Many examples of these ideas may be found in traditional books such as that by Hall and Knight [8], who discuss the matter at length.

The solutions to many combinatorial problems are polynomials. In some cases experimental solutions to these problems may be obtained for small values of the variable involved. If the assumption is made that a polynomial fits these solutions then the general form of solution may be predicted using the difference table and the ideas discussed above. Sawyer [14] gives an interesting discussion and a number of simple examples.

A harder example is discussed at length by Polya [12]. The problem is to determine the maximum number of regions into which three-dimensional space may be divided by n planes.

        1 plane produces 2 regions;
        2 planes can produce 4 regions;
        3 planes can produce 8 regions;
but 4 planes can only produce 15 regions at the most!

Polya's whole discussion of the problem, especially the geometrical significance of the difference table, is highly illuminating and should be consulted.

Exercise 1.13

Assuming that a polynomial fits the above data on planes and regions, and that you have been given enough information to find it, construct the difference table and show that the polynomial is

$$f(n) = \tfrac{1}{6}(n^3 + 5n + 6).$$

This function arises as
$$f(n) = f_0(n) + f_1(n) + f_2(n) + f_3(n).$$
Show that the maximum number of regions into which a plane may be divided by n lines is
$$f_0(n) + f_1(n) + f_2(n).$$
Suggest a generalization.

Exercise 1.14

A still more surprising sequence arises when a circle is subdivided by chords joining points on the circumference.

With 2 points the chord joining them produces 2 regions;
with 3 points the chords joining them produce 4 regions;
with 4 points the chords joining them produce 8 regions;
with 5 points the maximum number of regions is 16;
but with 6 points the maximum number of regions is 31.

This is a very good school example, which serves as a warning against guesswork unsupported by subsequent analysis and verification. Construct the difference table and show that the polynomial which takes the values 2, 4, 8, 16, 31 for $n = 2, 3, 4, 5, 6$, is
$$f(n) = f_0(n) + f_2(n) + f_4(n).$$
It is another matter to prove that this formula really does give the maximum number of regions which can arise from the chords joining n points, and this is left to the reader. The correspondence in *Mathematics Teaching* [6] may be consulted.

Exercise 1.15

Interesting complications arise in the difference tables of geometric sequences. For example, we have

$$\begin{array}{ccccccc} 1 & 2 & 4 & 8 & 16 & 32 & 64 \quad \dots \\ & 1 & 2 & 4 & 8 & 16 & 32 \quad \dots \\ & & 1 & 2 & 4 & 8 & 16 \quad \dots \\ & & & 1 & 2 & 4 & 8 \quad \dots \end{array}$$

This suggests the result
$$2^x = f_0(x) + f_1(x) + f_2(x) + f_3(x) + \dots .$$
Investigate relations of this kind.

1.5 Difference equations and differential equations

The Fibonacci sequence 1, 2, 3, 5, 8, 13, 21, ... is well known. It is defined by the relation
$$u_{n+2} = u_{n+1} + u_n, \quad n \geqslant 0, \qquad (1.2)$$
with $u_0 = 1$ and $u_1 = 2$.

What can be said about all sequences which satisfy Eq. (1.2), leaving aside the initial conditions? Before attempting to find any general algebraic expressions for such sequences it is useful to make two observations. Equation (1.2) is linear and homogeneous, and this shows that the sum of any two sequences of the type, adding term by term, is another sequence of the type. Also, if the terms of any sequence satisfying Eq. (1.2) are all multiplied by the same scalar multiplier the resulting sequence again satisfies the equation.

The situation is just as in Section 1.1, the sequences forming a vector space of dimension two. The two initial terms serve as co-ordinates, specifying a particular sequence in the space. The sequence with $u_0 = 1$ and $u_1 = 0$ is

$$1, 0, 1, 1, 2, 3, 5, \ldots,$$

and the sequence with $u_0 = 0$ and $u_1 = 1$ is

$$0, 1, 1, 2, 3, 5, 8, \ldots.$$

Any other sequence satisfying Eq. (1.2) is expressible as a linear sum of these two. Once again there is nothing special about these two sequences, any other pair would do (unless it was a pair in which one sequence was merely a numerical multiple of the other).

Similar remarks apply to any linear difference equation of the second order with constant coefficients, that is any equation of the form $u_{n+2} = au_{n+1} + bu_n$. The initial example on arithmetic progressions was merely the particular case concerned with sequences satisfying the relation

$$u_{n+2} = 2u_{n+1} - u_n, \quad n \geqslant 0.$$

The further techniques for solving equations of this type will not concern us; but they are readily available in algebra texts.

Similar remarks apply to homogeneous linear differential equations with constant coefficients—that is equations of the form

$$a\frac{d^2y}{dx^2} + b\frac{dy}{dx} + cy = 0, \tag{1.3}$$

where a, b and c are constant. Once again we will not be concerned with the techniques of solving such equations. These may be found in calculus books. Our concern is merely to note that even before any solutions have been found it is apparent that the sum of two solutions will give a further solution, and a scalar multiple of a solution will give a further solution also. That is to say the solutions of the equation form a linear space. In this context we may speak of a linear space or a vector space or a function space; the terms mean the same and there are only shades of emphasis or colloquial usage in employing the different ones.

It is proved in books on analysis that the dimension of the solution space of Eq. (1.3) is two.

In many of the applications of differential equations boundary conditions occur. Homogeneous boundary conditions, such as, for example,

$$
\begin{aligned}
&\text{(i)} \quad y(0) = 0,\\
\text{or} \quad &\text{(ii)} \quad y(4) = 0,\\
\text{or} \quad &\text{(iii)} \quad y'(0) = 0,\\
\text{or} \quad &\text{(iv)} \quad y'(7) = 0,
\end{aligned} \quad (1.4)
$$

select a subspace of the previous solution space. This is to say the sum of any two solutions of Eqs. (1.3) and (1.4) is again a solution, and a scalar multiple of any solution is again a solution. Under these more stringent conditions the dimension of the solution spaces is only one.

Note that boundary conditions of the type $y(0) = 3$ (say) or $y'(1) = 12$ do *not* lead to a subspace of solutions. (Why?) Neither do boundary conditions such as $y''(0) = 6$ (say). (Why?)

Function spaces also occur in connection with partial differential equations. Potential theory, as it occurs in the theory of gravitation or electricity and magnetism, provides examples. Consider the potential V in a two-dimensional space (x, y). This two-dimensional space is *physical* space, or at least physicists' idealized model of physical space. As we proceed we will be concerned with function spaces of various dimensions; these must not be confused with the (x, y) space.

A potential function is any twice differentiable function $V(x, y)$ satisfying

$$\frac{\partial^2 V}{\partial x^2} + \frac{\partial^2 V}{\partial y^2} = 0.$$

Thus

$V = $ constant is a potential function;

$V = x$ and $V = y$ are potential functions;

$V = xy$ is a potential function.

Before continuing we may note that potential functions form a linear space. (Why?) Can any further homogeneous potential functions of degree two be found? If $ax^2 + by^2$ is a potential function then it is necessary to have $a + b = 0$. Hence $x^2 - y^2$ is a potential function.

Exercise 1.16
Verify that *all* potential functions of degree two are of the form $\alpha xy + \beta(x^2 - y^2)$.

What potential functions can be found of degree three?
If
$$V = ax^3 + bx^2y + cxy^2 + dy^3$$
then
$$\frac{\partial^2 V}{\partial x^2} + \frac{\partial^2 V}{\partial y^2} = 6ax + 2by + 2cx + 6dy$$
$$= (6a + 2c)x + (2b + 6d)y.$$

The right-hand side of this equation is zero for all x and y if, and only if
$$6a + 2c = 2b + 6d = 0.$$
Hence
$$c = -3a \text{ and } b = -3d.$$

This gives the potential functions $x^3 - 3xy^2$ and $y^3 - 3x^2y$. All other potential functions of degree three can be shown to be linear combinations of these two.

These results may be summarized by observing that potential functions of degree one form a linear space of dimension two, functions of degree two and degree three do the same. Similar working shows that potential functions of degree n form a space of dimension two.

Similar results can be found for potential functions in (physical) three-dimensional space and the ideas can be used whenever potential theory is applied to scientific problems.

Exercise 1.17
Find some potential functions of degree 4 in (x, y) space.

Exercise 1.18
Find some potential functions of various degrees in (x, y, z) space.

1.6 Linear dependence in physics and chemistry

Physicists can frequently make use of a method called *dimensional analysis*. The method may be illustrated by using it to derive Mersenne's laws for the frequency of a vibrating string. (These examples may be omitted by those unfamiliar with the physics involved.)

EXAMPLE 1.7
If it is assumed on physical grounds that the law relating the frequency of a vibrating string to the tension, the mass per unit length and the length is of the form

$$\text{frequency} = (\text{tension})^\alpha \, (\text{line density})^\beta \, (\text{length})^\gamma \times \text{constant},$$

where α, β and γ are exponents which the method has to find, then we may proceed as follows.

Every physical quantity is expressible uniquely in terms of the fundamental units of mass, length and time. From previous physical knowledge frequency has dimensions T^{-1}, tension dimensions MLT^{-2}, mass per unit length dimensions ML^{-1}, and length dimensions L. Hence the law may be expressed in terms of the dimensions as

$$T^{-1} = (MLT^{-2})^\alpha (ML^{-1})^\beta L^\gamma. \qquad (1.5)$$

From the exponents of M, L and T respectively

$$0 = \alpha + \beta,$$
$$0 = \alpha - \beta + \gamma,$$
$$-1 = -2\alpha.$$

Hence $\alpha = \frac{1}{2}$, $\beta = -\frac{1}{2}$, and $\gamma = -1$. These three results are tantamount to what the physicist knows as Mersenne's laws.

It can be seen that the argument is really a vector argument. M may be represented by the vector $(1, 0, 0)$, L by the vector $(0, 1, 0)$ and T by the vector $(0, 0, 1)$. Equation (1.5) may then be rewritten as

$$(0, 0, -1) = \alpha(1, 1, -2) + \beta(1, -1, 0) + \gamma(0, 1, 0).$$

From this vector equation the three separate equations for α, β, and γ come just as before.

EXAMPLE 1.8

Kepler's third law of planetary motion may be deduced in the following way. Assuming that

planet's year = (radius of orbit)$^\alpha$ (mass of sun)$^\beta$
(gravitational constant)$^\gamma$ × constant,

then

$$(0, 0, 1) = \alpha(0, 1, 0) + \beta(1, 0, 0) + \gamma(-1, 3, -2).$$

The only difficult step here is knowing that the gravitational constant itself has dimensions, which are $M^{-1}L^3T^{-2}$. Then

$$0 = \beta - \gamma,$$
$$0 = \alpha + 3\gamma,$$
$$1 = -2\gamma.$$

Hence

$$\alpha = \tfrac{3}{2}, \qquad \beta = -\tfrac{1}{2} \text{ and } \gamma = -\tfrac{1}{2}.$$

This gives Kepler's third law, which includes the information that the square of the planet's year is proportional to the cube of its distance from the sun.

EXAMPLE 1.9

Assuming that the flow of liquid through a tube, in litres per second, depends on the radius of the tube, the pressure gradient in the tube, and the viscosity of the liquid, deduce the form of the law. We assume that

rate of flow = (radius)$^\alpha$ (pressure gradient)$^\beta$ (viscosity)$^\gamma$ × constant.

Then
$$(0, 3, -1) = \alpha(0, 1, 0) + \beta(1, -2, -2) + \gamma(1, -1, -1).$$

Hence
$$0 = \beta + \gamma,$$
$$3 = \alpha - 2\beta - \gamma,$$
and
$$-1 = -2\beta - \gamma.$$
Hence $\alpha = 4$, $\beta = 1$ and $\gamma = -1$.

EXAMPLE 1.10

When water flows from a reservoir through a V-shaped opening of angle θ it may be assumed that

rate of flow = (height of surface above bottom of V)$^\alpha g^\beta$ × constant,

where g is the acceleration due to gravity. There is also an obvious, and one would expect a complicated, dependence on the angle θ; but this being a dimensionless parameter it does not come into the calculation. This makes the method easy to apply, but it also means that the method is quite unable to deduce how the angle affects the flow.

In the usual way
$$(0, 3, -1) = \alpha(0, 1, 0) + \beta(0, 1, -2).$$

Hence $\alpha = \frac{5}{2}$ and $\beta = \frac{1}{2}$. Even though we do not know the dependence on θ it is an interesting conclusion that the rate of flow is proportional to the $\frac{5}{2}$ power of the height above the bottom of the opening. This is the basis of a practical method of determining the flow of rivers. A weir with a V-shaped notch may be calibrated to give the flow.

We have only sketched the method of dimensions in outline, and many questions arise as to its validity. All that is claimed here is that the method enables plausible physical conjectures to be made with a very small amount of calculation. There are some interesting biological applications in Maynard Smith [13].

From a mathematical point of view the method amounts to establishing a linear dependence relation between vectors in a vector space. The dependence of the new physical quantity on previously known quantities is determined, the exponents in the physical law becoming the coefficients in the vector equations. Similar ideas are at work when chemical equations are balanced. In most examples in school chemistry the balancing is no more than a matter of common sense. Thus, confronted with the reaction between common salt and sulphuric acid

$$NaCl + H_2SO_4 \rightarrow Na_2SO_4 + HCl$$

it is almost immediately apparent that in order to balance the equation must be written

$$2NaCl + H_2SO_4 \rightarrow Na_2SO_4 + 2HCl.$$

A more complicated example may show in greater detail what is involved. Consider the chemical reaction

$$Ca + H_3PO_4 \rightarrow Ca_3P_2O_8 + H_2.$$

As it stands this does not balance. The problem is to attach coefficients α, β, γ and δ so that

$$\alpha Ca + \beta H_3PO_4 \rightarrow \gamma Ca_3P_2O_8 + \delta H_2$$

does balance. Balance is achieved by ensuring that the same number of atoms of each element occurs each side. Therefore to balance the calcium requires

$$\alpha = 3\gamma.$$

To balance the hydrogen requires

$$3\beta = 2\delta.$$

To balance the phosphorus requires

$$\beta = 2\gamma,$$

and to balance the oxygen requires

$$4\beta = 8\gamma.$$

A solution to these equations is given by

$$\alpha = 3, \qquad \beta = 2, \qquad \gamma = 1, \qquad \delta = 3.$$

Any numbers proportional to these satisfy the equations, but the chemist takes the smallest solution in whole numbers.

The solution above has been presented in the form of simultaneous equations, but it can be seen as a matter of securing a linear dependence relation between vectors if we realize that the problem is to choose α, β, γ and δ so that

$$\alpha(1, 0, 0, 0) + \beta(0, 3, 1, 4) = \gamma(3, 0, 2, 8) + \delta(0, 2, 0, 0).$$

Exercise 1.19

The chemist who is convinced that the whole question is too simple to merit any mathematical consideration might try to balance the equations of the following reactions:

$$KMnO_4 + H_2SO_4 + KBr \longrightarrow K_2SO_4 + Br_2 + MnSO_4 + H_2O,$$
$$As_2S_3 + H_2O + HNO_3 \longrightarrow NO + H_3AsO_4 + H_2SO_4.$$
$$Fe_7S_8 + O_2 \longrightarrow Fe_3O_4 + SO_2.$$

If a systematic method is sought for dealing with the difficult cases then it is easily seen that we are concerned once again with finding the coefficients in linear dependence relations between vectors. Each chemical compound is to be regarded as a vector, and the base vectors in the space usually correspond to the elements in the reaction. However, in some chemical reactions radicals such as CH or NH_4 persist unchanged, and it may be convenient to regard these as 'base vectors'. This happens with the PO_4 radical in the example worked earlier, and as a result the algebra takes place not in a four-dimensional vector space but in a three-dimensional subspace.

It is not the concern of this review to develop algorithms in detail but to classify problems and to show many varied appearances of the basic ideas. This example differs from the previous ones in that chemists usually restrict their numerical coefficients to whole numbers, whereas in the previous examples there was the understanding that the coefficients could be fractions also. Putting it more formally, in the other examples the coefficients were taken from the *field* of rational numbers, whereas in this one they are taken from the *ring* of integers. This being so the present system should strictly be classified as a *module* rather than a vector space. The theory of modules is part of the theory of groups, and we will not pursue it any further; the interest here is in noting yet one more part of school mathematics in which the idea of linear dependence is important.

This leads very conveniently to the next point. In the earlier examples we worked, as we have just said, with the tacit understanding that the scalar multipliers were rational numbers, or perhaps real numbers. In either case they come from some number field, and this is the important thing. There are other linear spaces in which the scalar multipliers come from other less common fields, but the abstract theory and also the practical manipulations follow essentially the same patterns. This is because the scalar multipliers are always manipulated by the four operations of elementary arithmetic, and these proceed the same way in any field. The differences between different number fields are irrelevant when we are performing the calculations which concern us at any particular time.

1.7 Vector spaces over GF(2)

This identity in underlying structure is shown very clearly by considering some linear spaces in which the coefficients come from the number field consisting of only two elements 0 and 1, with addition and multiplication carried out modulo 2 by the tables

+	0	1		×	0	1
0	0	1	and	0	0	0
1	1	0		1	0	1

This number field is sometimes called GF(2), GF standing for *Galois field*, and sometimes called the field of residues modulo 2. This number field is frequently of theoretical interest because it provides very simple test cases. A generalization about *all fields* is false if it does not apply to this one. This field provides very simple numerical exercises. The surprising thing is that a number of situations which were by no means contrived to provide exercises on vector spaces in the first place turn out to be vector spaces in which the scalar multipliers are taken from GF(2).

The game of Nim

The game of Nim is discussed in various books of puzzles and mathematical diversions as well as in some books on number theory [9, 11]. It is a game for two players, in which there are three piles of matches from which the players draw alternately. When a draw is made any number of matches may be taken from one pile only. There are two variants of the game, in one the aim is to play in such a way as to take the last match, in the other the aim is to force the opponent to take the last match. It can be shown that the appropriate strategy in the two forms of the game is the same except that in the second form of the game the strategy has to be modified at the very end. This being so, we need consider only the first form.

A typical situation confronting a player is to be faced with piles containing, say, 13, 7 and 5 matches. It can be shown that the following method gives a winning strategy. First express the numbers in the binary scale, that is express them as 1101, 111 and 101, but think of these symbols not only as binary numbers but also as vectors, that is as $(1, 1, 0, 1)$, $(0, 1, 1, 1)$ and $(0, 1, 0, 1)$. Now comes the surprise.

The way to play is to draw matches from one of the piles in such a way as to make the three vectors linearly dependent! If they are linearly dependent already your opponent has a stranglehold if he knows how to use it. To apply the rule we see that in this situation the modification must be made to the first pile, because the numbers in piles can only be reduced and never increased. The move is to turn the vector $(1, 1, 0, 1)$ into $(0, 0, 1, 0)$; this is working to the rule because

$$(0, 0, 1, 0) = (0, 1, 1, 1) + (0, 1, 0, 1), \quad \text{(modulo 2)}.$$

This means reduce the pile with 13 matches to a pile of 2, leaving the opponent faced with 2, 7 and 5.

Whatever the opponent does now we must apply the rule again. Thus if he leaves 2, 6, 5 we must consider the vectors $(0, 0, 1, 0)$, $(0, 1, 1, 0)$ and $(0, 1, 0, 1)$, or what comes to the same thing in this context $(0, 1, 0)$, $(1, 1, 0)$ and $(1, 0, 1)$. We can set up linear dependence whilst at the same time reducing one of the piles by reducing the 5 to 4, that is by changing the vector $(1, 0, 1)$ to $(1, 0, 0)$; since

$$(1, 0, 0) = (0, 1, 0) + (1, 1, 0), \quad \text{(modulo 2)}.$$

The play is to leave 2, 6 and 4.

Different situations arise again according to the opponent's next move, but experiment with the situation will confirm that if the rule is applied after each of the opponent's moves then he will eventually lose. Why is this?

The soundness of the strategy is not difficult to establish, but we have space only to outline the argument. We may classify positions as 'safe' or 'unsafe' according as the three related vectors are dependent or independent.

It is quite easy to see that any move from a safe position leads to an unsafe one, since matches may only be taken from one pile and this is bound to upset the linear dependence relation. Therefore, if you are able to leave a safe position your opponent is forced to leave an unsafe one.

The next stage in the argument is a little harder. It is necessary to show that an unsafe position may always be changed to a safe one by withdrawing matches from one of the piles only. A little consideration shows how to do this.

Consider the *first* components of the vectors. If there is an even number of ones altogether in these three positions then proceed to the second component, and so on. When you have located the left-most position in which there is an odd number of ones this will tell you on which pile to operate. If there is only one then you must operate on the corresponding pile. Thus, faced with

$$(0, 1, 0, 1) \quad \text{or} \quad 5,$$
$$(0, 0, 1, 1) \quad \text{or} \quad 3,$$
$$(1, 0, 1, 1) \quad \text{or} \quad 11,$$

it is essential to draw from the third pile, so as to leave

$$
\begin{array}{ll}
(0, 1, 0, 1), & 5, \\
(0, 0, 1, 1), & 3, \\
(0, 1, 1, 0), & 6.
\end{array}
$$

But given

$$
\begin{array}{ll}
(1, 0, 0, 1), & 9, \\
(1, 1, 0, 0), & 12, \\
(1, 1, 1, 0), & 14,
\end{array}
$$

the situation is a little more complicated. The three ones in the first column tell us that to start with an adjustment must be made there, but adjustments have to be made to the third and fourth columns of the table as well. There are three ways of doing this; the 9 may be reduced to 2, the 12 may be reduced to 7, or the 14 may be reduced to 5, leaving respectively

$$
\begin{array}{lllllll}
(0, 0, 1, 0), & 2; & (1, 0, 0, 1), & 9; & & (1, 0, 0, 1), & 9; \\
(1, 1, 0, 0), & 12; & (0, 1, 1, 1), & 7; & \text{and} & (1, 1, 0, 0), & 12; \\
(1, 1, 1, 0), & 14; & (1, 1, 1, 0), & 14; & & (0, 1, 0, 1), & 5;
\end{array}
$$

and all of these are safe. It is difficult to put this process into words in such a way as to cover all eventualities, but it could be a splendid exercise in flow charting. It was one of the first demonstration examples employed to show the general public the ability of computers.

The game is eventually won by a player taking the last match, or matches, leaving nothing in any of the piles, and this position is safe. Therefore if a player who knows the strategy is confronted at any stage of the game by an unsafe position then he can leave a safe position after every move he makes and eventually win the game.

A fuller discussion of Nim and some similar games is given by O'Beirne [11]; see also Hardy and Wright [9].

The Hamming Code

The previous example showed how the selection of a strategic move in a game was a matter of establishing a linear dependence relation between vectors in what was at first sight a rather strange vector space. Similar vector spaces occur in practical contexts in the design of experiments and in the error-correcting codes which may be used in systems which transmit coded information.

It is well known that information is often stored in computers in some form of binary code, that is a code employing noughts and ones which may be represented simply in a variety of electrical or mechanical forms. When information is moved around inside machines or transmitted to other machines it may be very important to ensure that it is not impaired by errors arising from malfunctioning components or from other random sources. Provided that not too many errors are involved there are various ways in which they may be corrected automatically. One such method was devised by Hamming. The method is most simply explained by considering the case of a binary code in which the digits are grouped in fours, that is to say the code contains 16 basic characters which may be written:

$$0000, 0001, 0010, 0011, \ldots, 1111.$$

Obviously if one digit in any four-digit group is altered the character will be mistaken for another one.

To overcome this the four-digit characters are expanded to seven-digit characters as follows. The four digits are inserted in positions 3, 5, 6 and 7 in the new characters, and we may call them X_3, X_5, X_6 and X_7. X_1, X_2 and X_4 are then chosen so that

$$X_4 + X_5 + X_6 + X_7 = 0,$$
$$X_2 + X_3 + X_6 + X_7 = 0, \quad \text{(modulo 2)}.$$
$$X_1 + X_3 + X_5 + X_7 = 0.$$

Thus the character 1001 first becomes ..1.001, and when X_1, X_2 and X_4 have been calculated it becomes 0011001.

Of the 128 possible seven-digit characters only 16 occur as properly formed characters, and they are recognizable as correct because they satisfy the three checking equations above. When any other characters arise they indicate an error of some kind. These errors can be traced in a very elegant way. Suppose 0011001 becomes altered to 0011011, an error arising in the sixth place. When the checks are applied we now get

$$1 + 0 + 1 + 1 = 1,$$
$$0 + 1 + 1 + 1 = 1, \quad \text{(modulo 2)}$$
$$0 + 1 + 0 + 1 = 0.$$

But the three digits on the right-hand side of these equations are 110, and they give the position of the digit in which the error has arisen in binary notation. It is interesting to investigate this and to verify that, provided there is only one error, the error is always revealed in this way. If there are more errors than one then this method is powerless, but provided that the density of the errors is not too great it gives a way to correct errors automatically, because comparatively simple circuitry can be designed to detect the error and carry out the necessary change.

Some teaching uses of this code have been described in Reference [5]; here we will merely look at some of the ways in which vector spaces are involved.

The 128 seven-digit symbols may be regarded as a seven-dimensional vector space, if vectors are added in the sense that

$$1001101 + 0011011 = 1010110,$$

where we do not put commas between the components of the vector and where the addition of components is carried out modulo 2.

The 16 error-free seven-digit symbols form a subspace. (Check that the vector sum of two error-free characters is again an error-free character.) What is the dimension of this subspace? It is four, corresponding to the dimension of the original space of four-digit symbols. What is a basis for this error-free subspace of the seven-digit space? It is natural to take 1110000, 1001100, 0101010 and 1101001.

The various transformations carried out are mappings which may be shown diagramatically as in Fig. 1.3.

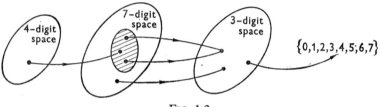

Fig. 1.3

The first mapping injects the four-digit space into a subspace of the seven-digit space. The checking process maps the seven-digit space on to a three-digit space (the space of triples of checking digits), and in this mapping the error-free seven-digit characters are mapped on to (0, 0, 0). Finally the triples of checking digits are mapped on the set $\{0, 1, \ldots 7\}$ by the rule $(\xi_1, \xi_2, \xi_3) \to 4\xi_1 + 2\xi_2 + \xi_3$ to give the position of the error.

The point of this example is merely to show that the language of vector spaces is flexible enough to describe situations such as this, where at first sight it might appear that quite new concepts would be involved.

Networks

Spaces of a similar kind occur in network theory. In some school work dealing with the application of Kirchhoff's laws to electrical networks the student is expected to handle the topological aspects of the network intuitively, for they are quite simple in the examples which he encounters. But when the student does this he is working intuitively in a particular kind of vector space. The electrical aspects of the problem are irrelevant as far as we are concerned, and we may look merely at the topology, which means in this case, looking at the way certain *nodes* are connected by *links*.

In figure 1.4 nodes A, B, C and D are connected by links a, b, c, d, e, f and g. We could agree on what we mean by saying that *cde*, *bdef* and *ab* (for example) form 'closed circuits'. How many different closed circuits can you find? The commonsense answer might be 12, but we can see that there might be an advantage in reconsidering exactly what constitutes a 'closed circuit' if closed circuits can be defined in such a way that they conform to an already familiar mathematical pattern. Still proceeding intuitively there seems to be some sort of sense whereby the closed circuit *bdef* can be regarded as made up of the loops *bcf* and *dec*, with the *c* 'cancelling' as the loops are added. This hunch can be formulated with mathematical precision if we consider the appropriate vector space, representing link *a* by the vector (1, 0, 0, 0, 0, 0, 0), and so on for each link up to *g*, which is represented by the vector (0, 0, 0, 0, 0, 0, 1). We adopt the convention that addition of vectors is to be done modulo 2. What then is a vector of the space such as (0, 1, 1, 0, 0, 1, 0)? It represents a set of links, in this case *b*, *c*, *f*.

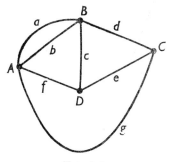

Fig. 1.4

What helps with the main problem is that closed loops form a subspace. Thus when we say *bcf* and *dec* combine to give *bdef*, we are saying that

$$(0, 1, 1, 0, 0, 1, 0) + (0, 0, 1, 1, 1, 0, 0) = (0, 1, 0, 1, 1, 1, 0).$$

What is the dimension of this subspace of closed loops? In this case it is 4; four suitable base vectors in vector notation are:

$$(1, 1, 0, 0, 0, 0, 0), \quad (0, 1, 1, 0, 0, 1, 0),$$
$$(0, 0, 1, 1, 1, 0, 0), \quad (0, 0, 0, 0, 1, 1, 1),$$

or in the other notation *ab*, *bcf*, *cde*, *efg*. Observe that if these conventions are adopted the objects of interest, that is the closed circuits, fit together into a familiar mathematical pattern, this being a vector space which is a subspace of a larger space. But this advantage is purchased at a certain price, for we have to agree that *abfeg* is a 'closed circuit' (and an electrician might not want to). We have to agree in more complex diagrams than Fig. 1.4 that the sum of two disjoint closed circuits is a 'closed circuit' and we have to include the 'empty circuit', which is the circuit containing no links at all, as a 'closed circuit'. Conventions of this kind are frequently the price which the practical man has to pay for mathematization.

It is no coincidence that the dimension of the subspace of closed circuits in Fig. 1.4 is four, which is the same as the number of separate regions in the plane which are bounded by the links of the network.

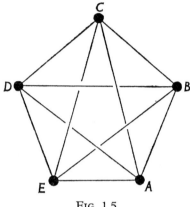

Fig. 1.5

It can be shown that this equality always holds for networks which can be drawn in a plane without the links crossing over. Not all networks can be so drawn, however much we pull the wires about in our efforts. For example, the network in Fig. 1.5 cannot be drawn flat in a plane.

Exercise 1.20

Investigate the network in Fig. 1.5 The dimension of the 'link space' is 10, but what is the dimension of the closed circuit subspace, and what is a suitable set of base circuits for it?

In more general circumstances there is a subspace of the closed circuit subspace which is of interest. This has no direct application in electricity but is interesting to explore as a piece of intuitive geometry.

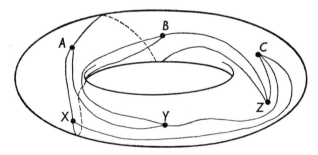

Fig. 1.6

Whilst networks of a certain degree of complexity cannot be drawn flat in a plane they can be drawn in more complicated surfaces; for example, some networks which cannot be drawn in a plane can be drawn on a torus. Fig. 1.6 shows such a case, this being a solution to the celebrated puzzle of providing houses A, B and C with gas, water and electricity services, X, Y and Z, without the connecting lines crossing. It is well known that this cannot be done in a plane.

Now in circumstances such as this some closed circuits bound an area of the torus and others do not. This can be seen in Fig. 1.7.

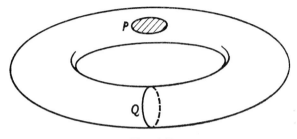

FIG. 1.7

Curve P bounds the shaded area of the torus, whereas curve Q does not bound any area in this way, and neither does any closed curve which circuits the torus longitudinally. It is not difficult to see that, with suitable conventions, the bounding circuits in a network form a subspace of the closed circuit subspace. This we have seen is, in its turn, a subspace of what we called the link space. These three types of spaces are a starting point for the study of homology, which is the branch of topology which deals with the connectedness of networks and the surfaces into which they will fit. Our investigations have involved, at most, three dimensions, and then only in the sense that curved surfaces such as the torus were considered as being immersed in three-space. The methods of homology are general methods in n dimensions.

It should also be mentioned that whilst we have employed the language of vector spaces, vector spaces are commutative groups as far as addition is concerned, and vector spaces over GF(2) are most frequently studied from the group point of view, and group language is employed.

There are yet further vector spaces associated with the electrical networks which prompted this discussion. For example, reverting to Fig. 1.4 we may ask: 'what sets of links must I cut to separate one or more nodes from the rest?' The question may be put another way. The nodes may be seen as little marbles and the links as pieces of elastic. 'If I pick up some marbles in one hand and the rest in the other what pieces of elastic will be stretched between my two hands?' Picking up A and B in one hand and C and D in the other, c, d, f and g would be stretched between the hands. *d*, *e*, and *g* could be stretched between the hands by picking the nodes up another way, but in no circumstances could just *a* and *c* be stretched between the hands.

Exercise 1.21
Explore this further, finding out the elements in the 'cut-set space', its dimension and its relation to the closed-circuit space described earlier. The same may be done for Fig. 1.5. More details may be found in Fletcher [5]; for fuller information consult Guillemin [7].

These particular vector spaces, or commutative groups, depending only on the language used to describe them, also occur in the practical problems of experimental design of the kind that arise in the biological sciences and in industrial quality control. A very readable introductory account may be found in Finney [4], and full details of industrial applications are given by Davies [3]. Some statistical experiments of this kind would be well within the scope of schools.

1.8 Spaces of infinite dimension

In the spaces we have explored so far there have been two operations. These have been called vector addition and scalar multiplication, and with their aid it has been possible to express other vectors in the space in terms of vectors which we have singled out as a basis. The precise meaning of addition and scalar multiplication depended on the context, but the methods of calculation were the same in the different contexts, and this was the interesting feature. When bases were chosen there was nothing unique about them, other sets of vectors would have done equally well for the purpose, but frequently one choice seemed more natural than others because of the particular circumstances of the problem.

More careful study of the general theory demonstrates what we have tended to take for granted on empirical evidence—that if the sets of base vectors are independent sets then the number of base vectors in each set is the same. This unique number is called the dimension of the space (or subspace) being considered, and we have seen many examples of this.

In many spaces which are of practical importance it emerges that it is not possible to find a *finite* basis. The very first example illustrated this. Arithmetical sequences form a space of dimension two, but this is a subspace of the space of all possible sequences; and it is not possible to select a finite number of sequences in terms of which all other sequences may be expressed linearly. The space of all possible sequences is not of finite dimension.

To some extent it is because of spaces such as this that present nomenclature is so varied. Some writers are inclined to speak of *vector spaces* when they are dealing with spaces which are of finite dimension, and to speak of *linear spaces* or *linear manifolds* when they are including spaces which are not. The only reason for introducing a variety of names into this book is to prepare the reader for reading elsewhere. Nomenclature is not fully standardized in modern algebra—which is still a developing subject—and it would be a great pity if this should discourage beginners. It is always essential to treat variations in nomenclature with caution, and to try to penetrate beyond the names to the ideas.

Even when spaces are not of finite dimension the idea of a basis is still important as there may still be some sense in which an infinite basis can be found. Thus in the case of the space of all sequences there is clearly a sense in which the particular vectors

$$(1, 0, 0, 0, \ldots),$$
$$(0, 1, 0, 0, \ldots),$$
$$(0, 0, 1, 0, \ldots),$$
$$(0, 0, 0, 1, \ldots), \quad \text{etc.}$$

form a basis. If we go on to consider the set of all real-valued functions, defined on some given interval of the real line, these will form a space, since any two can be added and any function can be multiplied by a scalar, but any attempt to pick out a particular set of functions as a basis in terms of which all real valued functions may be expressed linearly is doomed to failure. This clearly cannot be done with any finite set of functions, and those familiar with the notions of *denumerability* and *non-denumerability* will know that no denumerable set of functions, that is a set which can be listed as the natural numbers can be listed, will do so either. However, if suitably restricted subsets of real-valued functions are taken then it is possible to express them as infinite series of various kinds. This amounts to expressing them in terms of a basis which is denumerably infinite. Functions expressible as power series or as Fourier series form spaces of this kind. This kind of space is needed to provide a rigorous theoretical background to the problems discussed in Chapters 3 to 5. Indeed for some theoretical purposes it is preferable to go further and consider spaces of complex-valued functions.

One final example may indicate still further the enormous variety of mathematical systems which may be classified as linear spaces. A set of points in a plane is said to be *convex* if every point on the line segment joining any two members of the set is also a member of the set. An elementary discussion of convexity may be found in Reference [5]. It is possible to define the operation of adding two sets $\mathscr{X}$ and $\mathscr{Y}$ by taking some fixed point O in the plane as origin, and forming the set of points $\mathscr{Z}$, where $\mathscr{Z}$ is set of all points of the form

$$\mathbf{z} = \mathbf{x} + \mathbf{y}, \quad \text{with } \mathbf{x} \text{ in } \mathscr{X} \text{ and } \mathbf{y} \text{ in } \mathscr{Y}.$$

(**x** and **y** are added as vectors in the usual way).

It is not difficult to show that if $\mathscr{X}$ and $\mathscr{Y}$ are convex then $\mathscr{Z}$ is convex as well, and so the set of convex sets is closed under addition. Using origin O there is an obvious definition of the set $k\mathscr{X}$, where k is a scalar multiplier, and so we have both the operations necessary for a vector space. With the operations of addition and scalar multiplication defined the set of all convex sets in a plane is a vector space. It is clearly not possible to choose a finite basis for this space.

1.9 Definitions

We have deliberately chosen to introduce linear spaces in an exploratory, informal way. In consequence no formal definition has been given so far. The reader who requires the full details of the theory may find them in a book such as Mirsky [10], which gives the following definition of a linear manifold (space):

> Let F be a field and M a set of elements $X, Y, Z, \ldots$. Suppose that with each α in F and each X in M is associated a definite element of M, denoted by αX, and that with each pair X, Y in M is associated a definite element of M denoted by $X + Y$. Suppose further that these operations satisfy the following conditions for all X, Y, Z in M and all α, β in F:
>
> (i) $X + Y = Y + X$;
> (ii) $X + (Y + Z) = (X + Y) + Z$;
> (iii) the equation $Y + U = X$ is soluble for U;
> (iv) $(\alpha + \beta)X = \alpha X + \beta X$;
> (v) $\alpha(X + Y) = \alpha X + \alpha Y$;
> (vi) $(\alpha\beta)X = \alpha(\beta X)$;
> (vii) $1X = X$.

Then M is called a *linear manifold* over F.

Textbooks differ in the terminology they use for base vectors and the related ideas, but it is fairly standardized practice to say that a set of vectors form a *basis*, or a *base*, or a *minimal base* for a space when

(i) they are linearly independent,
(ii) every vector in the space can be expressed in terms of them.

It is desirable also to be able to say that a set of vectors *generate* or *span a space*. This means that they satisfy requirement (ii), but they are not necessarily linearly independent. In these cases it is always possible to find a smaller set of vectors which is linearly independent and which is a basis for the space, although it may not be practically convenient to do so in some circumstances.

Given any set of vectors $\{x_1, x_2, \ldots x_k\}$, they generate or span the space created by considering all possible linear sums of the form

$$\lambda_1 x_1 + \lambda_2 x_2 + \ldots + \lambda_k x_k,$$

where the λs are taken from some number field—usually the field of real numbers. Without further information we cannot be certain about the dimension of this space, because the xs might have linear dependence relations connecting them. We do know, of course, that the dimension of this vector space cannot exceed k.

References

1. BIRKHOFF, G. and MACLANE, S., *A Survey of Modern Algebra*, Macmillan (1953).
2. BOTSCH, O., *Praxis der Mathematik*, March 1966, pp. 76–7.
3. DAVIES, O. L., *The Design and Analysis of Industrial Experiments*, Oliver and Boyd (1956).
4. FINNEY, D. J., 'Statistical Science and Agricultural Research', *Mathl. Gaz.* 31, No. 293, pp. 21–30 (Feb. 1947).
5. FLETCHER, T. J., (ed.) *Some Lessons in Mathematics*, CUP (1964).
6. GLENN, J., 'The Quest for the Lost Region', *Mathematics Teaching*, 43, pp. 23–25 (1968); G. Matthews, Letter in *Mathematics Teaching*, 44, p. 9 (1968); R. Sibson, Letter in *Mathematics Teaching*, 45, p. 10 (1968).
7. GUILLEMIN, E. A., *The Mathematics of Circuit Analysis*, Wiley (1949).
8. HALL, H. S. and KNIGHT, S. R., *Algebra*, Macmillan (1887).
9. HARDY, G. H. and WRIGHT, E. M., *An Introduction to the Theory of Numbers*, OUP (1945).
10. MIRSKY, L., *An Introduction to Linear Algebra*, OUP (1955).
11. O'BEIRNE, T. H., *Puzzles and Paradoxes*, OUP (1965).
12. POLYA, G., *Induction and Analogy in Mathematics*, OUP (1954).
13. MAYNARD SMITH, J., *Mathematical Ideas in Biology*, CUP (1968).
14. SAWYER, W. W., *The Search for Pattern*, Pelican (1970).

2

Some Aspects of Geometry

It is well known that the introduction of co-ordinate methods into geometry, usually attributed to Descartes, was a major advance in the subject. It is a common view that the object of this was to be able to use algebraic methods to solve geometrical problems; but important as this is, it is interesting that Descartes himself seems to have been even more concerned with using geometrical configurations to illustrate and to clarify algebraic relations. The latter point of view is the one to be taken here.

If geometry is regarded as the science of physical space then it follows that any algebraic geometry whatever is applied mathematics; but in this book we are taking a much more limited view and we are concerned with aspects of geometry and algebra which are applicable to other branches of mathematics, such as mechanics and statistics, which by common agreement are 'applied mathematics'. It thus comes about that almost all of the geometry covered in this chapter is dealt with because it is needed later in the book in some non-geometrical context. In particular, almost all of this geometry is required in the statistics chapter—and much is required in the mechanics chapters also; not in the obvious sense that mechanics takes place in three-dimensional physical space, but for the less obvious reason that n-dimensional geometry is useful when explaining the behaviour of mechanical systems which have n independent moving parts. In order to establish the main ideas it is essential to illustrate them as we progress by numerous small-scale examples whose relevance it may be difficult to see if they are considered in isolation. For this reason the reader who wishes to reach the applications quickly is encouraged to skip this chapter and refer back later as necessary.

2.1 A note on matrix multiplication

It will be assumed that the reader has encountered the elementary ideas of multiplying two matrices together. This is explained in numerous school texts: References [2, 3, 5, 6, 7].

The essential idea is that multiplication is by row and column. Thus

$$\begin{bmatrix} a & b \\ c & d \end{bmatrix} \begin{bmatrix} p & q \\ r & s \end{bmatrix} = \begin{bmatrix} ap+br & aq+bs \\ cp+dr & cq+ds \end{bmatrix}.$$

The element in row 2 and column 1 of the product (for example) is derived from row 2 of the first matrix on the left-hand side and column 1 of the second matrix. The row $[c, d]$ and the column $\begin{bmatrix} p \\ r \end{bmatrix}$ are combined to give $cp + dr$. This process of combining two vectors, in this case a row vector and a column vector, by multiplying corresponding elements and then adding is called forming the *inner product*.

It is possible to multiply matrices which are not square; but for matrices **A** and **B** to be *conformable for the product* **AB** it is essential that the number of columns in **A** is the same as the number of rows in **B**, otherwise the inner products cannot be calculated. If **A** has r rows and s columns, and **B** s rows and t columns, then **AB** has r rows and t columns.

EXAMPLE 2.1

$$\begin{bmatrix} 1 \\ -4 \end{bmatrix} [2 \ \ 3] = \begin{bmatrix} 2 & 3 \\ -8 & -12 \end{bmatrix},$$

but

$$[2 \ \ 3] \begin{bmatrix} 1 \\ -4 \end{bmatrix} = [-10].$$

$$\begin{bmatrix} 1 & 0 \\ -1 & 1 \end{bmatrix} \begin{bmatrix} 7 \\ 5 \end{bmatrix} = \begin{bmatrix} 7 \\ -2 \end{bmatrix},$$

and it is not possible to form the product

$$\begin{bmatrix} 7 \\ 5 \end{bmatrix} \begin{bmatrix} 1 & 0 \\ -1 & 1 \end{bmatrix}.$$

However,

$$[7 \ \ 5] \begin{bmatrix} 1 & 0 \\ -1 & 1 \end{bmatrix} = [2 \ \ 5].$$

A square matrix with ones in the principal diagonal and zeros everywhere else is called a unit matrix and is usually denoted by **I**. (The principal diagonal is the diagonal extending from the top left to the bottom right-hand corners.) It is easy to show that it is always the case that **AI** = **IA** = **A**; where **I** is of such a size as to be conformable for the product with **A**.

2.2 Transformations

Today many school courses contain exercises on matrix transformations, usually restricted to two dimensions. We will now consider some of these, in two dimensions and in three dimensions as well. It will be necessary to assume a slight previous knowledge of three-dimensional co-ordinate geometry. In each example we will take note of any points which the transformation leaves invariant, and also of the inverse transformation where there is one. We will be concerned for the moment with *point* transformations, that is to say there is a fixed system of co-ordinates and the transformation moves points off to new addresses in this fixed co-ordinate system. The other point of view, that of a *co-ordinate* transformation, in which every point stays where it is but has its house renumbered, is also important, but this will be considered later.

When employing matrix methods in plane geometry it is useful to call the Cartesian co-ordinates (x_1, x_2) and to represent a point in the algebra by a column vector. With this understanding the standard notation for a transformation is $\mathbf{x} \rightarrow \mathbf{Ax}$. In detail we may write

$$\mathbf{x} = \begin{bmatrix} x_1 \\ x_2 \end{bmatrix}, \quad \mathbf{A} = \begin{bmatrix} a_{11} & a_{12} \\ a_{21} & a_{22} \end{bmatrix},$$

and so

$$\begin{bmatrix} x_1 \\ x_2 \end{bmatrix} \rightarrow \begin{bmatrix} a_{11}x_1 + a_{12}x_2 \\ a_{21}x_1 + a_{22}x_2 \end{bmatrix}.$$

This means that the point located at (x_1, x_2) is moved to the location $(a_{11}x_1 + a_{12}x_2, a_{21}x_1 + a_{22}x_2)$.

The most instructive examples are quite simple ones.

EXAMPLE 2.2

$$\mathbf{A} = \begin{bmatrix} 1 & 0 \\ 0 & -1 \end{bmatrix}.$$

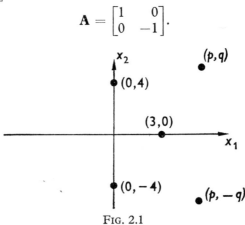

FIG. 2.1

The point $(3, 0)$ is left at $(3, 0)$.
The point $(0, 4)$ goes to $(0, -4)$.
The point (p, q) goes to $(p, -q)$.

The transformation is a reflection in the x_1-axis. Every point on this axis is left invariant, and only these points are left invariant.

This transformation is its own inverse; which means that if we perform the transformation a second time then every point is moved back to its original location.

EXAMPLE 2.3

$$\mathbf{A} = \begin{bmatrix} -1 & 0 \\ 0 & -1 \end{bmatrix}.$$

The point (p, q) goes to $(-p, -q)$.

The transformation is a half-turn about the origin. (It may also be described as a reflection in the origin.) The origin is the only invariant point, and the transformation is its own inverse.

EXAMPLE 2.4

$$\mathbf{A} = \begin{bmatrix} 0 & -1 \\ 1 & 0 \end{bmatrix},$$

Since $\begin{bmatrix} 0 & -1 \\ 1 & 0 \end{bmatrix} \begin{bmatrix} 2 \\ 0 \end{bmatrix} = \begin{bmatrix} 0 \\ 2 \end{bmatrix},$

the point $(2, 0)$ goes to $(0, 2)$,
the point $(0, 3)$ goes to $(-3, 0)$,
the point (p, q) goes to $(-q, p)$.

The transformation is a rotation about the origin through a right angle in the counter-clockwise direction.

The origin is the only invariant point, and the inverse transformation, a rotation about the origin through a right angle in the clockwise sense, has matrix

$$\begin{bmatrix} 0 & 1 \\ -1 & 0 \end{bmatrix}.$$

EXAMPLE 2.5

$$\mathbf{A} = \begin{bmatrix} 1 & 0 \\ 0 & 0 \end{bmatrix}.$$

The point $(2, 0)$ stays where it is,
the point $(2, 3)$ goes to $(2, 0)$.

The transformation projects every point orthogonally on to the x_1-axis. There is no inverse transformation in this case. The transformation superimposes a number of previously distinct points and after this it is impossible, by means of any linear algebraic transformation, to separate them again.

All the points on the x_1-axis are left invariant by the transformation.

EXAMPLE 2.6
$$\mathbf{A} = \begin{bmatrix} 2 & 0 \\ 0 & 0 \end{bmatrix}.$$

This is very like the previous case, but with the difference that points on the x_1-axis have their distance from the origin doubled. The origin is the only invariant point of the transformation, but the x-axis is an invariant line. This means that each point on the line is mapped somewhere on the line. The line itself is unchanged by the transformation although the individual points are displaced on it.

Again there is no inverse transformation.

EXAMPLE 2.7
$$\mathbf{A} = \begin{bmatrix} 1 & 1 \\ 1 & 1 \end{bmatrix}.$$

This again is a 'collapse', but of a more complicated kind,

the point (2, 3) goes to (5, 5),
the point (4, 1) goes to (5, 5) also,
the point (p, q) goes to $(p + q, p + q)$.

The origin is the only invariant point, but the line $x_1 = x_2$ is an invariant line. There is no inverse transformation.

EXAMPLE 2.8
$$\mathbf{A} = \begin{bmatrix} 2 & 0 \\ 0 & 2 \end{bmatrix}.$$

This is a stretch. The distance of every point from the origin is doubled,

the origin is the only invariant point,
every line through the origin is an invariant line,
the inverse transformation has a matrix

$$\begin{bmatrix} \tfrac{1}{2} & 0 \\ 0 & \tfrac{1}{2} \end{bmatrix}.$$

EXAMPLE 2.9
$$\mathbf{A} = \begin{bmatrix} 2 & 0 \\ 0 & 3 \end{bmatrix}.$$

This is also a stretch, but not an isotropic one; that is to say the magnitude of the stretch is different in different directions.

The origin is the only invariant point, but the x_1-axis and the x_2-axis are both invariant lines.

The inverse transformation has the matrix

$$\begin{bmatrix} \frac{1}{2} & 0 \\ 0 & \frac{1}{3} \end{bmatrix}.$$

EXAMPLE 2.10

$$\mathbf{A} = \begin{bmatrix} 1 & 1 \\ 0 & 1 \end{bmatrix}.$$

The point (p, q) is moved to $(p + q, q)$.

The transformation is a shear, in which all points on the x_1-axis are invariant and in which other points are, as it were, slid in layers; the amount of slide is proportional to the distance from the x_1-axis.

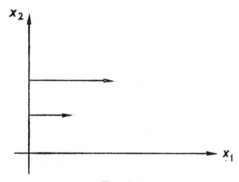

FIG. 2.2

The inverse transformation has the matrix

$$\begin{bmatrix} 1 & -1 \\ 0 & 1 \end{bmatrix}.$$

Many more illustrative examples could be given, but nearly all of them are more complicated combinations of types of transformation which have been considered above.

When transformations are performed successively the matrix of the resulting transformation is obtained by multiplying together the matrices of the separate transformations, taking care to arrange them in the correct order.

Exercise 2.1

In each of the above examples consider the effect of repeating the transformations a number of times. Which transformations produce closed cycles? Which transformations have no additional effect if they are repeated?

Exercise 2.2
Construct similar examples in three dimensions. An interesting variety of cases can arise, and it is worth pursuing this at some length.

Later examples will cover cases where the matrix of the transformation is rectangular.

2.3 The solution of simultaneous equations

In the previous section the inverse matrix was accepted intuitively as the matrix of the transformation which 'undid' a previous transformation. The only matrices which have inverses in this way are square, but not all square matrices possess inverses. If a matrix **A** has an inverse, which is usually denoted by $\mathbf{A}^{-1}$, then it has the property

$$\mathbf{A}\mathbf{A}^{-1} = \mathbf{A}^{-1}\mathbf{A} = \mathbf{I},$$

where **I** is the unit matrix of the appropriate order. This raises the important questions, given a square matrix **A**, how are we to decide if it has an inverse, and how may the inverse be calculated?

The calculation of inverse matrices is intimately connected with the solution of simultaneous equations. Indeed, the two processes are essentially the same, and it is very misleading to students if inverse matrices are presented to them as something that will take the hard work out of solving simultaneous equations, because in all cases of any importance the steps involved in calculating the inverse matrix correspond exactly to the steps involved in solving the associated simultaneous equations by successive elimination.

Simultaneous equations can be solved by determinants, and inverse matrices can be introduced by way of determinants and their minors. Whilst this is useful background knowledge for students, it must be emphasized that these methods are usually very wasteful in time and effort in practical calculation.

We will now consider the solution of simultaneous equations by successive elimination, and see how this leads very naturally to all the main ideas about inverse matrices. It is helpful to consider three equations in three unknowns.

EXAMPLE 2.11
Consider, as an example, the equations

$$\begin{aligned} x_1 - 2x_2 - 3x_3 &= 3, \\ 2x_1 - x_2 - 4x_3 &= 7, \\ 3x_1 - 3x_2 - 5x_3 &= 8. \end{aligned} \qquad (2.1)$$

THE SOLUTION OF SIMULTANEOUS EQUATIONS

These equations may be written in matrix form, as

$$\begin{bmatrix} 1 & -2 & -3 \\ 2 & -1 & -4 \\ 3 & -3 & -5 \end{bmatrix} \begin{bmatrix} x_1 \\ x_2 \\ x_3 \end{bmatrix} = \begin{bmatrix} 3 \\ 7 \\ 8 \end{bmatrix}. \tag{2.2}$$

Two matrices are said to be equal if and only if they are equal element by element; so it comes about that the one matrix equation (2.2) is equivalent to the three scalar equations (2.1).

Using obvious abbreviations Eq. (2.2) may be written

$$\mathbf{Ax} = \mathbf{b}. \tag{2.3}$$

The long-term aim is to derive a matrix $\mathbf{A}^{-1}$, so that the solution of Eq. (2.3) can be carried out merely by a matrix multiplication by $\mathbf{A}^{-1}$; thus

$$\mathbf{A}^{-1}.\mathbf{Ax} = \mathbf{A}^{-1}\mathbf{b},$$
$$\mathbf{Ix} = \mathbf{A}^{-1}\mathbf{b},$$
$$\mathbf{x} = \mathbf{A}^{-1}\mathbf{b}.$$

The whole discussion is directed towards constructing the matrix $\mathbf{A}^{-1}$ with the desired properties.

If Eqs. (2.1) are solved by successive elimination the steps might be as follows. Subtracting suitable multiples of the first equation from the other two,

$$\begin{aligned} x_1 - 2x_2 - 3x_3 &= 3, \\ 3x_2 + 2x_3 &= 1, \\ 3x_2 + 4x_3 &= -1. \end{aligned} \tag{2.4}$$

Now subtracting the second equation from the third,

$$\begin{aligned} x_1 - 2x_2 - 3x_3 &= 3, \\ 3x_2 + 2x_3 &= 1, \\ 2x_3 &= -2. \end{aligned} \tag{2.5}$$

From this point the solution is easy, but it is worth considering the details. We may proceed systematically by dividing the third equation by two, and then subtracting appropriate multiples of it from the first two, getting

$$\begin{aligned} x_1 - 2x_2 &= 0, \\ 3x_2 &= 3, \\ x_3 &= -1. \end{aligned} \tag{2.6}$$

Following a system, we may now divide the second equation by three and add twice the result to the first equation. This produces the solution

$$\begin{aligned} x_1 &= 2, \\ x_2 &= 1, \\ x_3 &= -1. \end{aligned} \tag{2.7}$$

The next point to observe is that the steps taken are in no way dictated by the numbers appearing on the right-hand side of the equations. Instead of Eqs. (2.1) we might have more generally

$$\begin{aligned} x_1 - 2x_2 - 3x_3 &= y_1, \\ 2x_1 - x_2 - 4x_3 &= y_2, \\ 3x_1 - 3x_2 - 5x_3 &= y_3. \end{aligned} \qquad (2.8)$$

Proceeding step by step as before, we now get

$$\begin{aligned} x_1 - 2x_2 - 3x_3 &= y_1, \\ 3x_2 + 2x_3 &= -2y_1 + y_2, \\ 3x_2 + 4x_3 &= -3y_1 + y_3. \end{aligned} \qquad (2.9)$$

Then

$$\begin{aligned} x_1 - 2x_2 - 3x_3 &= y_1, \\ 3x_2 + 2x_3 &= -2y_1 + y_2, \\ 2x_3 &= -y_1 - y_2 + y_3; \end{aligned} \qquad (2.10)$$

then

$$\begin{aligned} x_1 - 2x_2 &= -\tfrac{1}{2}y_1 - \tfrac{3}{2}y_2 + \tfrac{3}{2}y_3, \\ 3x_2 &= -y_1 + 2y_2 - y_3, \\ x_3 &= -\tfrac{1}{2}y_1 - \tfrac{1}{2}y_2 + \tfrac{1}{2}y_3; \end{aligned} \qquad (2.11)$$

and finally

$$\begin{aligned} x_1 &= -\tfrac{7}{6}y_1 - \tfrac{1}{6}y_2 + \tfrac{5}{6}y_3, \\ x_2 &= -\tfrac{1}{3}y_1 + \tfrac{2}{3}y_2 - \tfrac{1}{3}y_3, \\ x_3 &= -\tfrac{1}{2}y_1 - \tfrac{1}{2}y_2 + \tfrac{1}{2}y_3. \end{aligned} \qquad (2.12)$$

As a check, substituting $y_1 = 3$, $y_2 = 7$, $y_3 = 8$ gives the original answer in equations (2.7).

There are now several points to observe. The calculation can be broken down into steps consisting of either

(i) multiplying an equation by a non-zero constant; or
(ii) adding a multiple of one equation to another.

(We can, of course, 'subtract' by 'adding' a negative multiple.)

To reduce the amount of printing in the calculation the steps were often carried out two, or even three, at a time, but the principle is that the calculation can be done by the successive performance of these 'elementary steps'. It is sometimes convenient in calculations, but not essential, to change the order in which the equations are written, and this can be carried out by using a third elementary step

(iii) interchanging a pair of equations.

The operations (i) and (ii) are the link with Chapter 1, where they were seen as the two basic operations in a vector space. In this type of calculation equations are handled as vectors. Operation (iii) is sometimes needed if the solutions are to be listed in the proper order. This process of solution will sometimes break down, because not all sets of simultaneous equations have a unique solution. These exceptional cases are illustrated by some of the following examples, but it will not be necessary for us to spend much time studying detailed techniques for these special cases.

Exercises

Apply the above method to the following sets of equations. In each case give two geometrical interpretations to what is happening.

(a) Interpret the equations as defining a transformation in which, against a fixed frame of co-ordinates, the point at address (x_1, x_2, x_3) is moved to the new address (y_1, y_2, y_3). In this case, solving the equations amounts to finding which points are mapped on to a given point.

(b) Interpret the equations as defining planes in the usual sense of three-dimensional Cartesian geometry, with co-ordinates (x_1, x_2, x_3), and interpret any peculiarities which the equations may have in terms of the relative situations of the planes.

Exercise 2.3

$$x_1 - 2x_2 - x_3 = 0,$$
$$2x_1 + x_2 + 5x_3 = 0,$$
$$3x_1 - 2x_2 + 3x_3 = 0.$$

The only solution of the equations is $x_1 = x_2 = x_3 = 0$.

(a) The transformation leaves the origin invariant; apart from this, without further work, we cannot say very much more about it.

(b) The equations represent three planes passing through the origin.

Exercise 2.4

$$x_1 - 2x_2 - x_3 = 0,$$
$$2x_1 + x_2 + 5x_3 = 3,$$
$$3x_1 - 2x_2 + 3x_3 = 2.$$

The solution of the equations is $x_1 = 3, x_2 = 2, x_3 = -1$.

(a) The transformation involved is exactly the same as in the previous question, and we have learnt that the point which is mapped on to $(0, 3, 2)$ is $(3, 2, -1)$.

(b) We have three planes, intersecting at $(3, 2, -1)$.

Exercise 2.5

$$2x_1 - x_2 + 3x_3 = 0,$$
$$x_1 + 2x_2 + x_3 = 0,$$
$$3x_1 - 4x_2 + 5x_3 = 0.$$

Whilst (0, 0, 0) is clearly a solution, do not jump to the conclusion that it is the only solution; any set of numbers in the ratio

$$x_1 : x_2 : x_3 = -7 : 1 : 5$$

is a solution. The basic method of solution has to be adapted to yield this.

(a) The transformation maps a whole line of points on to the origin. Since (second equation) + (third equation) = 2 × (first equation) the equations are not independent, and an image point with co-ordinates (y_1, y_2, y_3) satisfies the equation

$$y_2 + y_3 = 2y_1;$$

that is to say the transformation compresses the entire three-dimensional space into this plane.

(b) Using the other picture we have three planes through the origin which intersect in the line $x_1 : x_2 : x_3 = -7 : 1 : 5$.

Exercise 2.6

$$2x_1 - x_2 + 3x_3 = 2,$$
$$x_1 + 2x_2 + x_3 = 1,$$
$$3x_1 - 4x_2 + 5x_3 = 3.$$

The standard method of solution has to be adapted to cover this case once again, and we find that once again there is an infinity of solutions. These are most easily expressed in vector form; $(1, 0, 0)$ plus any multiple of $(-7, 1, 5)$ gives a solution, but this may be expressed in alternative forms.

(a) The map involved is exactly the same as in the previous question. A line of points as mapped on to (2, 1, 3).

(b) From the second point of view we have three planes which intersect in the line through (1, 0, 0) in the direction $(-7, 1, 5)$.

Exercise 2.7

$$2x_1 - x_2 + 3x_3 = 2,$$
$$x_1 + 2x_2 + x_3 = 1,$$
$$3x_1 - 4x_2 + 5x_3 = 4.$$

This example differs from the previous one only in having 4 instead of 3 on the right-hand side of the last equation. This makes a big difference, because there are now no solutions at all. (Why? and in what way does the standard procedure for solution shown in the text break down?)

(a) In the transformation no point is mapped on to (2, 1, 4).

(b) The three planes have no common point of intersection because they form a triangular prism.

Exercise 2.8

$$x_1 - 2x_2 + 3x_3 = 5,$$
$$2x_1 - 4x_2 + x_3 = 5,$$
$$3x_1 - 6x_2 + 2x_3 = 8.$$

With this set of equations there is no way of breaking up the combination of terms $x_1 - 2x_2$, and all that can be derived is

$$x_1 - 2x_2 = 2 \quad \text{and} \quad x_3 = 1.$$

From an alternative point of view:

(first equation) $= 5 \times$ (third equation) $- 7 \times$ (second equation)

and so the equations are not independent.

(a) All of the image points, naming them (y_1, y_2, y_3), lie in the planes $y_1 = 5y_3 - 7y_2$, and a whole line of points is mapped on to $(5, 5, 8)$, as indeed a whole line of points is mapped on to any point in this plane.

(b) From the other point of view the three planes involved intersect in a line.

From these examples it appears that breakdown of the standard procedure for solution is associated with linear dependence of the left-hand sides of the equations. It is immediately clear that unless the right-hand sides of the equations are linearly dependent in the same way the equations are inconsistent and there is no solution at all. If the same linear dependence relation holds on both sides of the equations then the effective number of equations is reduced; one equation, at the very least, may be seen as a consequence of the others, and then there is a multiplicity of solutions.

The situation in interpretation (b) above is then that the three planes instead of being in a 'general' position in space and intersecting in a unique point, are disposed in some 'special' way, intersecting in a line, forming the faces of a triangular prism, or being parallel (a case we have not previously considered) according to the particular circumstances.

In interpretation (a) the breakdown of the standard procedure for solution is associated with a *singular* or *degenerate* transformation which maps many points on to the same image point, and no points at all on to some possible image points. In discussing these ideas the notion of the *rank* of a transformation, or its associated matrix, is very useful. A matrix maps one vector space into another vector space, and the *rank* of a matrix may be defined as the dimension of its image space. Rank is a very important idea in the development of matrix theory, but as this book is intended to be an elementary introduction to applications we will proceed as far as possible without employing the notion of rank: this is considered further towards the end of Section 2.9.

We have referred rather loosely to 'the standard method of solution', but beyond giving numerical examples we have not attempted to define the procedure exactly. It is possible to define a suitable procedure exactly, and one way of doing so is to specify it by a flow-chart as is shown in Fig. 2.3. From a chart such as this a computer program may be

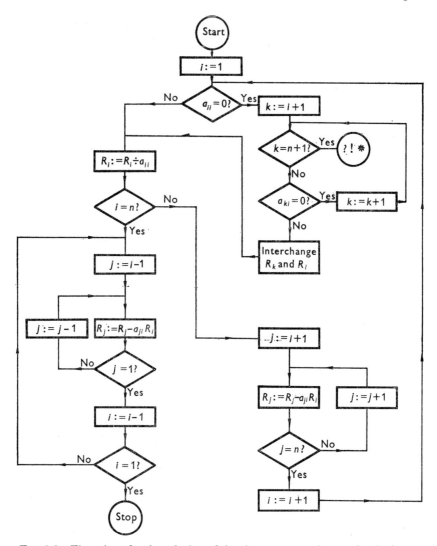

FIG. 2.3 Flow chart for the solution of simultaneous equations or for the inversion of matrices.

drawn up in some suitable language for the numerical solution of simultaneous linear equations. The flow chart may be regarded as specifying a machine into which the numbers appearing in the equations

$$\mathbf{Ax} = \mathbf{b}$$

are fed, after having been assembled into a matrix with n rows and $n + 1$ columns, which may be denoted by

$$[\mathbf{A} \mid \mathbf{b}].$$

THE SYSTEMATIC USE OF MATRICES

The machine then processes this matrix until it is reduced to the form

$$[\mathbf{I} \mid \mathbf{x}],$$

which has the required answers in the final column. Alternatively, if the case is a special one in which there is no unique solution, the machine sounds an alarm and a special investigation is required.

Henceforth, unless otherwise stated, we will assume that we are dealing with the solution of equations in the *non-singular* case, i.e. the case where there is a unique solution.

2.4 The systematic use of matrices

Throughout the long previous discussion matrices have hardly been mentioned. It is now time to show how they relate to what has been done. We return to Example 2.11. Associated with the whole process of solution there is an inverse process. Equations (2.8) may be regarded as equations in unknowns x_1, x_2, x_3 with 'knowns' y_1, y_2, y_3; whereas Eqs. (2.12) are equations in unknowns y_1, y_2, y_3 with knowns x_1, x_2, x_3. (2.12) is the solution of (2.8), but equally well (2.8) is the solution of (2.12). The reverse steps involved in going from (2.12) to (2.8) are all elementary steps as defined earlier.

We will now show that when the equations are written in matrix form the matrices appearing in Eqs. (2.8) and Eqs. (2.12) are inverses of one another. Earlier Eq. (2.1) was written in matrix form as Eq. (2.2) and then Eq. (2.3). This can clearly be done at each stage of the chain, Eqs. (2.8–2.12); but at first sight to do so appears to be only a notational device, with a new matrix summarizing the results of a step which has been taken by non-matrix means. If this were all, then matrices would not be very important; but it is not all, and we will shortly see that matrices are involved in the calculation in an altogether deeper way. Expressed in matrix notation the steps may be rewritten as:

$$\begin{bmatrix} 1 & -2 & -3 \\ 2 & -1 & -4 \\ 3 & -3 & -5 \end{bmatrix} \begin{bmatrix} x_1 \\ x_2 \\ x_3 \end{bmatrix} = \begin{bmatrix} 1 & 0 & 0 \\ 0 & 1 & 0 \\ 0 & 0 & 1 \end{bmatrix} \begin{bmatrix} y_1 \\ y_2 \\ y_3 \end{bmatrix}; \qquad (2.8)$$

$$\begin{bmatrix} 1 & -2 & -3 \\ 0 & 3 & 2 \\ 0 & 3 & 4 \end{bmatrix} \begin{bmatrix} x_1 \\ x_2 \\ x_3 \end{bmatrix} = \begin{bmatrix} 1 & 0 & 0 \\ -2 & 1 & 0 \\ -3 & 0 & 1 \end{bmatrix} \begin{bmatrix} y_1 \\ y_2 \\ y_3 \end{bmatrix}; \qquad (2.9)$$

$$\begin{bmatrix} 1 & -2 & -3 \\ 0 & 3 & 2 \\ 0 & 0 & 2 \end{bmatrix} \begin{bmatrix} x_1 \\ x_2 \\ x_3 \end{bmatrix} = \begin{bmatrix} 1 & 0 & 0 \\ -2 & 1 & 0 \\ -1 & -1 & 1 \end{bmatrix} \begin{bmatrix} y_1 \\ y_2 \\ y_3 \end{bmatrix}; \qquad (2.10)$$

$$\begin{bmatrix} 1 & -2 & 0 \\ 0 & 3 & 0 \\ 0 & 0 & 1 \end{bmatrix} \begin{bmatrix} x_1 \\ x_2 \\ x_3 \end{bmatrix} = \begin{bmatrix} -\tfrac{1}{2} & -\tfrac{3}{2} & \tfrac{3}{2} \\ -1 & 2 & -1 \\ -\tfrac{1}{2} & -\tfrac{1}{2} & \tfrac{1}{2} \end{bmatrix} \begin{bmatrix} y_1 \\ y_2 \\ y_3 \end{bmatrix} ; \quad (2.11)$$

$$\begin{bmatrix} 1 & 0 & 0 \\ 0 & 1 & 0 \\ 0 & 0 & 1 \end{bmatrix} \begin{bmatrix} x_1 \\ x_2 \\ x_3 \end{bmatrix} = \begin{bmatrix} -\tfrac{7}{6} & -\tfrac{1}{6} & \tfrac{5}{6} \\ -\tfrac{1}{3} & \tfrac{2}{3} & -\tfrac{1}{3} \\ -\tfrac{1}{2} & -\tfrac{1}{2} & \tfrac{1}{2} \end{bmatrix} \begin{bmatrix} y_1 \\ y_2 \\ y_3 \end{bmatrix} . \quad (2.12)$$

It will be remembered that some of the steps given here are in fact the combination of two or three elementary steps. At this moment this tabulated calculation is seen only as a summary, in matrix shorthand as it were, of the calculation which was done before. If it is examined, however, it may be seen that at each stage, indeed at each elementary step, exactly the same operations are carried out on the 3×3 matrices on the two sides of the equation. It is possible to elevate this to a theorem, which might be termed the *principle of row transformations*.

Following on directly from the elementary steps defined earlier, it is natural to define three types of elementary row transformation as follows:

   (i) the multiplication of a row by a non-zero constant;
   (ii) the addition of any multiple of some row to some other row;
   (iii) the interchange of a pair of rows.

It might be appropriate once again to stress that (i) and (ii) are merely the standard operations of vector spaces, the rows of the matrix being considered as vectors. Transformation (iii) merely enables us to relabel if we wish.

What we are temporarily calling the principle of row transformations might be stated as follows.

THEOREM 2.1
If, given any matrix equation **AB** = **CD**, *any elementary row transformation, or succession of transformations, is carried out on* **A** *and* **C**, *to produce* **A**★ *and* **C**★ *respectively, then* **A**★**B** = **C**★**D**.

Proof
This result is not difficult to prove directly from the definition of matrix multiplication, since this multiplication uses the *rows* of the first factor altogether, treating all the elements in a row the same way. The details of the proof are left as an exercise.

This principle can be useful in matrix computations, but we will be looking at things a little differently. Explanations can sometimes be clarified by introducing *elementary transformation matrices*. Any elementary row transformation can be carried out on a matrix by pre-multiplying it by a suitable matrix. This is most easily shown by examples. Verify that

$$\begin{bmatrix} 1 & 0 & 0 \\ 0 & 2 & 0 \\ 0 & 0 & 1 \end{bmatrix} \begin{bmatrix} a & b & c \\ l & m & n \\ p & q & r \end{bmatrix} = \begin{bmatrix} a & b & c \\ 2l & 2m & 2n \\ p & q & r \end{bmatrix},$$

$$\begin{bmatrix} 1 & 0 & 3 \\ 0 & 1 & 0 \\ 0 & 0 & 1 \end{bmatrix} \begin{bmatrix} a & b & c \\ l & m & n \\ p & q & r \end{bmatrix} = \begin{bmatrix} a+3p & b+3q & c+3r \\ l & m & n \\ p & q & r \end{bmatrix},$$

$$\begin{bmatrix} 1 & 0 & 0 \\ 0 & 0 & 1 \\ 0 & 1 & 0 \end{bmatrix} \begin{bmatrix} a & b & c \\ l & m & n \\ p & q & r \end{bmatrix} = \begin{bmatrix} a & b & c \\ p & q & r \\ l & m & n \end{bmatrix}.$$

These examples illustrate elementary row transformations of types (i), (ii) and (iii) respectively. The general rule is easily stated. Any row transformation may be carried out on a matrix by pre-multiplying it by the matrix $\mathbf{I}^\star$; where $\mathbf{I}^\star$ is derived from the unit matrix by carrying out the appropriate row transformations on it. The word 'elementary' is deliberately omitted from this enunciation, as the rule applies to combinations of elementary transformations as well as to single ones.

This rule is of great importance. A proof is really a matter of verification, applying the definitions to the various cases which arise. Alternatively, the rule may be regarded as a simple corollary or Theorem 2.1, putting $\mathbf{D} = \mathbf{A}$ and $\mathbf{B} = \mathbf{C} = \mathbf{I}$.

There is a similar theory for column transformations and post-multiplication, which may be verified in a similar way.

The passage from Eqs. (2.8) to (2.9) involved subtracting twice the first row from the second, and three times the first from the third; so the matrix operator required is obtained by performing these operations on the unit matrix, getting

$$\begin{bmatrix} 1 & 0 & 0 \\ -2 & 1 & 0 \\ -3 & 0 & 1 \end{bmatrix}.$$

Now verify that

$$\begin{bmatrix} 1 & 0 & 0 \\ -2 & 1 & 0 \\ -3 & 0 & 1 \end{bmatrix} \begin{bmatrix} 1 & -2 & -3 \\ 2 & -1 & -4 \\ 3 & -3 & -5 \end{bmatrix} = \begin{bmatrix} 1 & -2 & -3 \\ 0 & 3 & 2 \\ 0 & 3 & 4 \end{bmatrix}.$$

The verification that the multiplication on the right-hand side of Eqs. (2.8) and (2.9) is also correct is very easy this time.

Verify that the passage from Eq. (2.9) to Eq. (2.10) involves premultiplication by

$$\begin{bmatrix} 1 & 0 & 0 \\ 0 & 1 & 0 \\ 0 & -1 & 1 \end{bmatrix}.$$

Find the appropriate matrices in other cases for yourself.

After all of this preliminary work we can now identify inverse matrices and state their fundamental properties. There are shorter paths than the one we have taken, but our method brings out the connexion with the solution of simultaneous equations and, what is very important, indicates a feasible method of calculating inverse matrices. Recapitulating, from Eq. (2.8)

$$\mathbf{Ax} = \mathbf{Iy}, \qquad (2.8)$$

by a sequence of elementary steps, each one of which may be carried out by multiplying on the left by an appropriate transformation matrix, we deduce Eq. (2.12), which we may write

$$\mathbf{Ix} = \mathbf{By}; \qquad (2.12)$$

and furthermore the process is reversible.

Denoting the transformation matrices by $\mathbf{T}_1, \mathbf{T}_2, \ldots \mathbf{T}_m$ (say), and applying them in the correct order to Eq. (2.8),

$$\mathbf{T}_m \ldots \mathbf{T}_2 \mathbf{T}_1 \mathbf{Ax} = \mathbf{T}_m \ldots \mathbf{T}_2 \mathbf{T}_1 \mathbf{Iy}.$$

Putting $\mathbf{T}_m \ldots \mathbf{T}_2 \mathbf{T}_1 = \mathbf{T}$, this becomes

$$\mathbf{TAx} = \mathbf{TIy} = \mathbf{Ty}.$$

But this is simply Eq. (2.12), and so

$$\mathbf{TA} = \mathbf{I} \quad \text{and} \quad \mathbf{T} = \mathbf{B}.$$

Hence

$$\mathbf{BA} = \mathbf{I}. \qquad (2.13)$$

The argument may now be reversed, starting with Eq. (2.12) and working back. This enables us to prove that

$$\mathbf{AB} = \mathbf{I}. \qquad (2.14)$$

It is necessary to establish this because, in general, matrices do not commute, and we cannot assume that Eq. (2.14) is true merely because Eq. (2.13) is established.

The matrix $\mathbf{B}$ is called the inverse of $\mathbf{A}$, and is usually denoted by $\mathbf{A}^{-1}$. The property $\mathbf{AA}^{-1} = \mathbf{A}^{-1}\mathbf{A} = \mathbf{I}$ is so important that it is worth stating explicitly.

Note that the flow chart in Fig. 2.3, which was previously used to solve equations, can be used just as well to invert matrices. As a direct consequence of the process which we have just been discussing, if the matrix **A** and the unit matrix are assembled into a matrix

$$[\mathbf{A} \mid \mathbf{I}],$$

which has n rows and $2n$ columns, and this is then processed by the flow chart, at the end we produce

$$[\mathbf{I} \mid \mathbf{A}^{-1}].$$

If **A** does not have an inverse, the flow chart indicates 'breakdown' on the way.

Exercise 2.9
Use the flow chart on the matrices appearing in Examples 2.2–2.10 and find their inverses where possible.

Exercise 2.10
Prove that a matrix cannot have more than one inverse, i.e. that if $\mathbf{AB} = \mathbf{BA} = \mathbf{I}$ and $\mathbf{AC} = \mathbf{CA} = \mathbf{I}$, then necessarily $\mathbf{B} = \mathbf{C}$. In fact it is just as easy to prove a stronger result; given that $\mathbf{BA} = \mathbf{I}$ and $\mathbf{AC} = \mathbf{I}$ then necessarily $\mathbf{C} = \mathbf{B}$, and $\mathbf{AB} = \mathbf{I}$ also.

Exercise 2.11 (harder)
Given a square matrix **A** show that the inversion process breaks down if and only if there is a linear dependence relation between the rows.

Exercise 2.12
Deduce Eqs. (2.13) and (2.14) from Eqs. (2.8) and (2.12) by direct use of Theorem 2.1, without employing elementary transformation matrices.

At this stage it will be useful to introduce another piece of notation. **A**′ will be used to denote the matrix which is the *transpose* of **A**—that is the matrix whose rows are the columns of **A** and whose columns are the rows of **A**. Using this notation enables the column vector

$$\begin{bmatrix} x_1 \\ x_2 \\ x_3 \end{bmatrix}$$

to be printed more conveniently as $(x_1, x_2, x_3)'$.

There are certain standard theorems about transposed matrices which are proved in most texts. We will make little use of these, and so we merely state the results and leave the proofs as exercises for the reader.

Exercise 2.13

If **A** and **B** are two matrices which are conformable for the product **AB** prove that

$$(\mathbf{AB})' = \mathbf{B}'\mathbf{A}'.$$

Verify this in various numerical cases if you cannot give a general proof. Do not restrict yourself to square matrices.

Deduce that if **A** has an inverse then

$$(\mathbf{A}')^{-1} = (\mathbf{A}^{-1})'.$$

It follows from this that the flow chart may be used to invert matrices by columns just as well as by rows.

Prove also that

$$(\mathbf{ABC})' = \mathbf{C}'\mathbf{B}'\mathbf{A}'.$$

2.5 Eigenvectors

In Section 2.2 we investigated the invariant points and invariant lines associated with some simple 2×2 matrix transformations. It is now time to consider this problem more generally. When a transformation $\mathbf{x} \to \mathbf{Ax}$ is carried out it is clear that the origin is always an invariant point; the interest is therefore in asking what other invariant points there are. As a number of the earlier examples indicated, this question is rather too general. Only in the case of some very simple transformations are points other than the origin invariant, but in most of the cases investigated there were invariant lines, and these always passed through the origin.

The interesting question therefore becomes—given a matrix transformation $\mathbf{x} \to \mathbf{Ax}$, and interpreting the vector **x** as the Cartesian co-ordinates of a point in a space of suitable dimension, what lines through the origin are invariant under the transformation? If **x** is a vector associated with a point on a line of this kind then, since **Ax** is some other point in the same direction from the origin, the co-ordinates of this point are also given by $\lambda\mathbf{x}$, where λ is a scalar. The question becomes, which vectors **x** have associated with them a scalar λ such that

$$\mathbf{Ax} = \lambda\mathbf{x}? \qquad (2.15)$$

A vector **x** satisfying a relation of the type in Eq. (2.15) is called an *eigenvector* and λ is called the associated *eigenvalue*.

Note that inevitably there is a certain difficulty for the beginner in that the two new notions of eigenvalue and eigenvector are defined simultaneously—this is inherent in the origins of the problem.

An example in three dimensions should make the general procedures clear.

EXAMPLE 2.12

What are the eigenvectors and eigenvalues of the matrix

$$\mathbf{A} = \begin{bmatrix} -1 & 2 & 2 \\ -8 & 7 & 4 \\ -13 & 5 & 8 \end{bmatrix}?$$

Working with a fixed set of co-ordinate axes this sends the point represented by the vector $\begin{bmatrix} x_1 \\ x_2 \\ x_3 \end{bmatrix}$ to the new location represented by

$$\mathbf{Ax} = \begin{bmatrix} -x_1 + 2x_2 + 2x_3 \\ -8x_1 + 7x_2 + 4x_3 \\ -13x_1 + 5x_2 + 8x_3 \end{bmatrix}. \tag{2.16}$$

If the new co-ordinates are scalar multiples λ of the old, then

$$\begin{aligned} -x_1 + 2x_2 + 2x_3 &= \lambda x_1, \\ -8x_1 + 7x_2 + 4x_3 &= \lambda x_2, \\ -13x_1 + 5x_2 + 8x_3 &= \lambda x_3. \end{aligned} \tag{2.17}$$

Hence

$$\begin{aligned} (-1-\lambda)x_1 + \quad\ 2x_2 + \quad\ 2x_3 &= 0, \\ -8x_1 + (7-\lambda)x_2 + \quad\ 4x_3 &= 0, \\ -13x_1 + \quad\ 5x_2 + (8-\lambda)x_3 &= 0. \end{aligned} \tag{2.18}$$

In matrix form the last two sets of equations are

$$\mathbf{Ax} = \lambda \mathbf{x}, \tag{2.19}$$

and

$$(\mathbf{A} - \lambda \mathbf{I})\mathbf{x} = 0. \tag{2.20}$$

At this stage it is customary to require the student to know a standard theorem about linear equations and determinants; ways round this will be considered later. The theorem states that a set of linear homogeneous equations has a solution in which the unknowns are not all zero if and only if the determinant of the coefficients on the left-hand side is zero. (Homogeneous means that the terms on the right-hand side are all zero.)

This means that Eq. (2.18) has solutions other than $x_1 = x_2 = x_3 = 0$ (which is no concern to us) if and only if the determinant

$$\begin{bmatrix} -1-\lambda & 2 & 2 \\ -8 & 7-\lambda & 4 \\ -13 & 5 & 8-\lambda \end{bmatrix} = 0. \tag{2.21}$$

This can also be written as

$$|\mathbf{A} - \lambda \mathbf{I}| = 0. \tag{2.22}$$

The standard procedure is now to expand the determinant in Eq. (2.21), getting eventually

$$\lambda^3 - 14\lambda^2 + 63\lambda - 90 = 0. \tag{2.23}$$

There is no short cut in this calculation—quite a lot of elementary algebra is involved however it is done. This equation, which may be in the form shown in Eqs. (2.21), (2.22) or (2.23), is called the *characteristic equation* of matrix **A**.

It is now necessary to solve this equation, and in this case it is not difficult as numbers have been chosen which work out. It factorizes readily,

$$(\lambda - 3)(\lambda - 5)(\lambda - 6) = 0,$$

giving values of λ,

$$\lambda = 3, 5 \text{ or } 6.$$

These are the eigenvalues of **A**.

It must be noted that, in general, the calculation does not work out as easily as this, and these examples in which the algebra is relatively convenient are only illustrative. For practical numerical calculation other methods have to be employed, and these are introduced in Section 6.6. This being so, we may remark at this stage that a knowledge of determinants is certainly not essential to make the steps from Eq. (2.18) to Eq. (2.23). An alternative method is as follows.

Working with Eq. (2.18), subtract twice the first equation from the second, and $\frac{1}{2}(8 - \lambda)$ times the first from the third; this eliminates x_3, getting

$$(-6 + 2\lambda)x_1 + (3 - \lambda)x_2 = 0,$$
$$-\tfrac{1}{2}(18 - 7\lambda + \lambda^2)x_1 - (3 - \lambda)x_2 = 0.$$

Hence, using the two equations,

$$\frac{x_1}{x_2} = \frac{\lambda - 3}{2\lambda - 6} = \frac{\lambda - 3}{\tfrac{1}{2}(\lambda^2 - 7\lambda + 18)}.$$

Cross-multiplying and rearranging we obtain Eq. (2.23). Actually in this case the factor $(\lambda - 3)$ emerges in the course of the calculation, but this cannot be expected to happen in general. We need to take note that we have assumed that $x_2 \neq 0$, and this point may be checked later. It is always possible to circumvent the determinant form of the characteristic equation by a calculation such as this, but there are technical difficulties in stating the general theory using methods of this kind, and nearly every textbook which the student is likely to consult at this stage will call for some knowledge of determinants. However, all of the examples which follow can be solved without determinants, although some may be solved more quickly with them, and it may be of interest to give a general procedure in the form of a flow chart, Fig. 2.4, which enables the

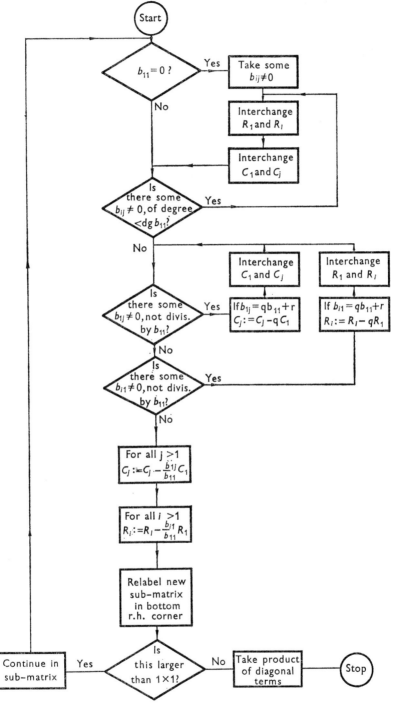

FIG. 2.4 Flow chart for the derivation of the characteristic equation of the matrix **A**. The input to the flow chart is the matrix $\mathbf{B} = \mathbf{A} - \lambda\mathbf{I}$.

characteristic equation in polynomial form to be calculated from the matrix $\mathbf{A} - \lambda \mathbf{I}$. This flow chart is not very convenient for longhand calculation, but it can be converted into a computer program quite easily.

We now resume the discussion from the point at which the eigenvalues $\lambda = 3, 5, 6$ have been found. To find the eigenvector corresponding to $\lambda = 3$ this value of λ has to be substituted into Eqs. (2.18), getting

$$-4x_1 + 2x_2 + 2x_3 = 0,$$
$$-8x_1 + 4x_2 + 4x_3 = 0,$$
$$-13x_1 + 5x_2 + 5x_3 = 0.$$

It is in the nature of the problem that the equations obtained at this stage are not independent, and also that x_1, x_2, x_3 can only be determined as far as their ratios. Elementary manipulation gives the solution of these equations as

$$x_1 : x_2 : x_3 = 0 : 1 : -1.$$

Thus the eigenvector corresponding to the eigenvalue $\lambda = 3$ is $(0, 1, -1)'$. It is worth checking that

$$\mathbf{A} \begin{bmatrix} 0 \\ 1 \\ -1 \end{bmatrix} = \begin{bmatrix} 0 \\ 3 \\ -3 \end{bmatrix}.$$

Note the use of the prime to make clear that $(0, 1, -1)$ has to be treated as a column vector.

A similar calculation now has to be carried out with $\lambda = 5$. Equations (2.18) now become

$$-6x_1 + 2x_2 + 2x_3 = 0,$$
$$-8x_1 + 2x_2 + 4x_3 = 0,$$
$$-13x_1 + 5x_2 + 3x_3 = 0.$$

Elementary manipulation, of which we do not give details, produces

$$x_1 : x_2 : x_3 = 1 : 2 : 1.$$

The eigenvector corresponding to the eigenvalue $\lambda = 5$ is therefore $(1, 2, 1)'$. Once more, check that

$$\mathbf{A} \begin{bmatrix} 1 \\ 2 \\ 1 \end{bmatrix} = \begin{bmatrix} 5 \\ 10 \\ 5 \end{bmatrix}.$$

When $\lambda = 6$ the equations giving the eigenvector become

$$-7x_1 + 2x_2 + 2x_3 = 0,$$
$$-8x_1 + x_2 + 4x_3 = 0,$$
$$-13x_1 + 5x_2 + 2x_3 = 0.$$

From this, after some working
$$x_1 : x_2 : x_3 = 2 : 4 : 3.$$
The eigenvector corresponding to $\lambda = 6$ is $(2, 4, 3)'$.

Check that
$$\mathbf{A} \begin{bmatrix} 2 \\ 4 \\ 3 \end{bmatrix} = \begin{bmatrix} 12 \\ 24 \\ 18 \end{bmatrix}.$$

These results may be summarized for subsequent reference. The matrix
$$\mathbf{A} = \begin{bmatrix} -1 & 2 & 2 \\ -8 & 7 & 4 \\ -13 & 5 & 8 \end{bmatrix}$$
has eigenvalues

$\lambda = 3$ with eigenvector $(0, 1, -1)'$;
$ = 5$ with eigenvector $(1, 2, 1)'$;
$ = 6$ with eigenvector $(2, 4, 3)'$.

When the matrix operates on vectors its effect on the eigenvectors is especially simple, it merely multiplies them by a scalar quantity. This means that for some purposes, as will be seen later, it is simpler to express vectors in terms of the eigenvectors rather than in terms of the 'obvious' base vectors $(1, 0, 0)'$, $(0, 1, 0)'$ and $(0, 0, 1)'$. This is one more case where the obvious basis for a vector space is not the most convenient when there is some specialized job to be done.

Exercise 2.14

Re-examine Examples 2.2–2.10 from this point of view. Some of these illustrate points of difficulty and we will comment on them briefly.

In Example 2.4
$$\mathbf{A} = \begin{bmatrix} 0 & -1 \\ 1 & 0 \end{bmatrix}.$$
$(x_1, x_2)'$ is an eigenvector if
$$-x_2 = \lambda x_1 \quad \text{and} \quad x_1 = \lambda x_2.$$
This requires
$$\lambda^2 x_2 = -x_2,$$
so either $x_2 = 0$, which means $x_1 = 0$ as well, and the solution is useless to us, or $\lambda^2 = -1$, which means $\lambda = \pm i$.

It may be surprising that we are prepared to consider complex eigenvalues, but a little calculation will verify that

$\lambda = +i$ has eigenvector $(1, -i)$,
$ = -i$ has eigenvector $(1, +i)$.

Just as complex numbers provide roots for algebraic equations which do not have them, or do not have the full number associated with their degree, in the real field, so the use of complex numbers enables eigenvalues and eigenvectors to be found for matrices which do not have them over the real field.

In Example 2.5

$$\mathbf{A} = \begin{bmatrix} 1 & 0 \\ 0 & 0 \end{bmatrix}.$$

$(x_1, x_2)'$ is an eigenvector if

$$x_1 = \lambda x_1, \qquad 0 = \lambda x_2.$$

This requires either $x_2 = 0$, in which case $\lambda = 1$ and x_1 is arbitrary; or $\lambda = 0$, in which case $x_1 = 0$ and x_2 is arbitrary.
Therefore,

$\lambda = 1$ is an eigenvalue with eigenvector $(1, 0)'$;
$\lambda = 0$ is an eigenvalue with eigenvector $(0, 1)'$.

In Example 2.10

$$\mathbf{A} = \begin{bmatrix} 1 & 1 \\ 0 & 1 \end{bmatrix}.$$

$(x_1, x_2)'$ is an eigenvector if

$$x_1 + x_2 = \lambda x_1, \qquad x_2 = \lambda x_2.$$

Unless x_1 and x_2 are both zero, a case of no interest to us, these equations demand $x_2 = 0$ and $\lambda = 1$. Hence $\lambda = 1$ is an eigenvalue, and the only eigenvalue, and the only eigenvector corresponding is $(1, 0)'$.

Verify that the following matrices have the eigenvalues and eigenvectors indicated. Some of these examples demand considerable numerical calculations if they are attempted without knowing the answers.

Exercise 2.15

$\begin{bmatrix} \cos 2\theta & \sin 2\theta \\ \sin 2\theta & -\cos 2\theta \end{bmatrix}$ $\quad \lambda = 1; \quad x_1 : x_2 = \cos\theta : \sin\theta$
$\quad\quad\quad\quad\quad\quad\quad\quad\quad\quad\lambda = -1; \quad x_1 : x_2 = \sin\theta : -\cos\theta.$

This matrix reflects the plane in the line $x_2 = x_1 \tan\theta$.

Exercise 2.16

$\begin{bmatrix} 0 & 7 & -6 \\ -1 & 4 & 0 \\ 0 & 2 & -2 \end{bmatrix}$ $\quad \lambda = 1; \quad (9, 3, 2)'.$
$\quad\quad\quad\quad\quad\quad\quad\lambda = 2; \quad (4, 2, 1)'.$
$\quad\quad\quad\quad\quad\quad\quad\lambda = -1; \quad (5, 1, 2)'.$

Exercise 2.17

$$\begin{bmatrix} 9 & -6 & 2 \\ -6 & 8 & -4 \\ 2 & -4 & 4 \end{bmatrix} \quad \begin{aligned} \lambda &= 1; & (1, 2, 2)'. \\ \lambda &= 4; & (2, 1, -2)'. \\ \lambda &= 16; & (2, -2, 1)'. \end{aligned}$$

Exercise 2.18

$$\begin{bmatrix} 0 & 0 & 0 & -1 \\ 1 & 0 & 0 & -4 \\ 0 & 1 & 0 & -6 \\ 0 & 0 & 1 & -4 \end{bmatrix}.$$

This is a very special case. The characteristic equation has a four-fold root, and there is only the one eigenvalue, $\lambda = -1$. It will be found that there is only one eigenvector associated with this eigenvalue, $(1, 3, 3, 1)'$.

Exercise 2.19

$$\begin{bmatrix} 2 & -2 & 3 \\ 10 & -4 & 5 \\ 5 & -4 & 6 \end{bmatrix}.$$

This introduces a similar type of complication.
$\lambda = 2$ is an eigenvalue with eigenvector $(4, 15, 10)'$.
$\lambda = 1$ is an eigenvalue which occurs as a double root in the characteristic equation, and there is only one corresponding eigenvector, $(1, 5, 3)'$.

Exercise 2.20

$$\begin{bmatrix} 7 & 4 & -1 \\ 4 & 7 & -1 \\ -4 & -4 & 4 \end{bmatrix}.$$

$\lambda = 12$ is an eigenvalue with eigenvector $(1, 1, -1)'$.
$\lambda = 3$ is an eigenvalue which occurs as a double root of the characteristic equation. The equations giving the eigenvector are

$$\begin{aligned} 4x_1 + 4x_2 - x_3 &= 0, \\ 4x_1 + 4x_2 - x_3 &= 0, \\ -4x_1 - 4x_2 + x_3 &= 0. \end{aligned}$$

This is a new situation; this time there is only one independent equation and so there is a multiplicity of solutions. One way of explaining it is to say that $(1, 0, 4)'$ and $(0, 1, 4)'$ are both eigenvectors, and so is any linear combination of them. $\lambda = 3$ has associated with it an eigenspace of dimension 2. This is an invariant plane in the transformation.

Exercise 2.21

$$\begin{bmatrix} 2 & 0 & 0 \\ 0 & 2 & 0 \\ 0 & 0 & 2 \end{bmatrix}.$$

Any non-zero vector is an eigenvector with eigenvalue 2. $\lambda = 2$ is a triple root of the characteristic equation and has associated with it an eigenspace of dimension 3.

Exercise 2.22

$$\begin{bmatrix} 5 & 1 & 1 \\ 1 & 5 & -1 \\ 1 & -1 & 5 \end{bmatrix}.$$

$\lambda = 3$ is an eigenvalue with eigenvector $(-1, 1, 1)'$.
$\lambda = 6$ is an eigenvalue, and the equations for the eigenvector are

$$-x_1 + x_2 + x_3 = 0,$$
$$x_1 - x_2 - x_3 = 0,$$
and
$$x_1 - x_2 - x_3 = 0.$$

Instead of having two independent equations there is only one equation,

$$x_1 - x_2 - x_3 = 0,$$

and any set of numbers satisfying this equation gives an eigenvector with eigenvalue 6. For example

$(1, 1, 0)',\quad (1, 0, 1)',\quad (1, -1, 2)',\quad (2, 1, 1)',\quad (0, 1, -1)'$

are all eigenvectors. The eigenvectors form a two-dimensional space; in other words, there is a plane of eigenvectors in this case, as in Exercise 2.20.

In general an $n \times n$ matrix has n eigenvalues and n eigenvectors. It turns out that it is not difficult to develop a general theory to cover these cases. However, as many of the examples bring out, there are exceptions to this; a matrix may *degenerate* by not having as many eigenvalues and eigenvectors as might be anticipated. It is hard to generalize about the dimension of the eigenspace associated with a repeated eigenvalue, and this creates difficulties when it comes to constructing a general theory. When multiple roots of the characteristic equation arise in applications or in problems they nearly always demand special care. There is, however, one piece of good fortune to note. When a matrix is symmetric a full complement of eigenvectors can always be found—but we will not prove that here. (Symmetric matrices are matrices with the defining property $\mathbf{A}' = \mathbf{A}$.)

2.6 The inner product and orthogonality

When geometrical problems are tackled by vector methods it is necessary to have techniques for computing angles between lines. This can be done as follows. The theorem of Pythagoras in three dimensions indicates that the vector $\mathbf{x} = (x_1' \ x_2' \ x_3)$ associated with the point with Cartesian co-ordinates (x_1, x_2, x_3) has length $\sqrt{(x_1^2 + x_2^2 + x_3^2)}$. This can be written in matrix shorthand as $\sqrt{(\mathbf{x'x})}$, because

$$\mathbf{x'x} = [x_1\ x_2\ x_3]\begin{bmatrix} x_1 \\ x_2 \\ x_3 \end{bmatrix} = x_1^2 + x_2^2 + x_3^2.$$

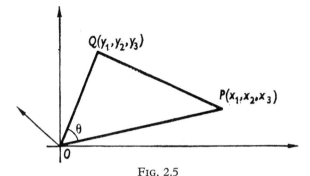

Fig. 2.5

If we consider a triangle POQ (Fig. 2.5) with P having co-ordinates (x_1, x_2, x_3) and Q having co-ordinates (y_1, y_2, y_3), and angle POQ equal to θ, then by the cosine rule for a triangle

$$PQ^2 = OP^2 + OQ^2 - 2OP.OQ \cos \theta.$$

Working with Cartesian co-ordinates, and using the theorem of Pythagoras to calculate PQ^2, OP^2 and OQ^2,

$$(x_1 - y_1)^2 + (x_2 - y_2)^2 + (x_3 - y_3)^2$$
$$= x_1^2 + x_2^2 + x_3^2 + y_1^2 + y_2^2 + y_3^2 - 2OP.OQ \cos \theta.$$

Hence $\quad OP.OQ \cos \theta = x_1 y_1 + x_2 y_2 + x_3 y_3.$

The expression on the right-hand side is called the inner product of (x_1, x_2, x_3) and (y_1, y_2, y_3), and it may be expressed as the matrix product $\mathbf{x'y}$ or $\mathbf{y'x}$.

From this very important formula a number of corollaries can be derived.

$$\cos \theta = \frac{x_1 y_1 + x_2 y_2 + x_3 y_3}{\sqrt{(x_1^2 + x_2^2 + x_3^2)} \sqrt{(y_1^2 + y_2^2 + y_3^2)}},$$

or in alternative notation

$$\cos \theta = \frac{\mathbf{x'y}}{\sqrt{(\mathbf{x'x})} \sqrt{(\mathbf{y'y})}}.$$

If $\mathbf{x}$ is a unit vector, i.e. $\mathbf{x'x} = 1$, then the projection of $\mathbf{y}$ in the direction $\mathbf{x}$ has magnitude $OQ \cos \theta$, and so it is equal to $OQ \cos \theta \times \mathbf{x}$, which is $\mathbf{x}(\mathbf{x}'\mathbf{y})$. For subsequent reference it will also be useful to record that the projection of $\mathbf{y}$ in the direction of a unit vector $\mathbf{x}$ has magnitude

$$\mathbf{x}'\mathbf{y} = \mathbf{y}'\mathbf{x} = x_1 y_1 + x_2 y_2 + x_3 y_3. \tag{2.24}$$

A further corollary is easily seen. Two non-zero vectors $\mathbf{x}$ and $\mathbf{y}$ are orthogonal (perpendicular) if and only if $\mathbf{x}'\mathbf{y} = 0$. Also if $\mathbf{x}$ and $\mathbf{y}$ are both unit vectors then θ, the angle between them, is given by

$$\cos \theta = \mathbf{x}'\mathbf{y}.$$

In two dimensions it is quite convenient to use the gradient of a line to describe its direction, but in more dimensions the direction of a line cannot be specified by means of a single parameter, and it is necessary to specify directions either by means of *direction cosines* or *direction ratios*. The direction cosines of a line are the cosines of the angles which it makes with the co-ordinate axes. (The reader may need to recall that it is standard mathematical practice to speak of the angle between two lines in space even if they do not intersect; in this case the angle may be measured by measuring the angle between any suitable pair of lines parallel respectively to the original two.) As an immediate result of the theorem of Pythagoras the sum of the squares of the direction cosines of any line is equal to one. Any set of numbers proportional to the direction cosines is called a set of direction ratios. A set of ratios can be converted to cosines by dividing by the square root of the sum of their squares. This is called *normalizing* the vector.

Orthogonality is an important idea in vector spaces. We have just seen that two vectors $\mathbf{x}$ and $\mathbf{y}$ are orthogonal if $\mathbf{x}'\mathbf{y} = 0$, or, what comes to the same thing, if $\mathbf{y}'\mathbf{x} = 0$. Orthogonality, or being perpendicular, is a familiar idea in elementary geometry, and we have just seen how the idea is incorporated into co-ordinate geometry. It is involved, however, in other vector space situations as well, and the next two exercises show some unusual appearances of the idea of orthogonality.

Exercise 2.23

Refer back to the networks in Section 1.7, in particular to Exercise 1.21. Verify that all the vectors in the cut-set space are orthogonal to all of the vectors in the closed-circuit space, when arithmetic is done modulo 2. Verify this also in other similar examples and seek a general reason for it.

Exercise 2.24

Most books which are published today bear a standard book number. This usually consists of nine digits, arranged in blocks of 3, 5 and 1 respectively. Few readers realize that the last digit is a check, calculated in the following way. Calling the first eight digits $d_2, \ldots, d_9$ (d_1 is an additional digit used in international standard book numbers), d_{10} is calculated by the formula

$$d_{10} = 2d_2 + 3d_3 + \ldots + 9d_9, \qquad \text{(modulo 11)}.$$

If d_{10} works out to be 10 it is written as X.

Show that this means that all standard book numbers are orthogonal to the vector (2, 3, 4, 5, 6, 7, 8, 9, 10) (modulo 11).

Orthogonality is often at work in error-detecting and error-correcting codes. Examples of this may be found in the Hamming code in Section 1.7.

Before leaving the topic of orthogonality we may mention an idea which is important in general theoretical developments, but as we do not intend to apply it in this book we will not spend time on details or prove any results. The idea is almost obvious in the elementary geometrical applications, but it is not quite so obvious in some other less-familiar vector spaces. It can be shown that in any vector space, once we have a method for calculating inner products, it is always possible to choose a set of base vectors which are mutually orthogonal. There is, furthermore, an algorithm which can be expressed as a flow chart which enables some arbitrary set of base vectors to be replaced by an orthogonal set of base vectors. This process is most commonly called the Schmidt process, but sometimes the names of Gram and Hilbert are linked with it as well.

The ideas of orthogonality, Pythagoras's theorem and projection have been explained in terms of the geometry of two and three dimensions, but it is clear that the formulation which we have been using is an algebraic formulation, and the corresponding algebraic expressions and formulae can be employed in vector spaces of any finite number of dimensions, and there is even hope of employing the ideas in spaces of infinite dimension if the definitions can be modified suitably, and if any problems of convergence which arise can be overcome. In Chapter 7 these ideas are required in n-dimensional space in order to handle certain problems in statistics.

2.7 Conics and quadrics

Remembering that it is the purpose of this chapter to assemble a relatively small number of powerful geometrical images for reference later in other contexts, the geometry attempted in this section will be very limited. Thus we will now study central loci of the second degree—central meaning that we include the ellipse and the hyperbola and the corresponding loci of higher dimensions but exclude the parabola and the paraboloids. Of these the ellipses and the ellipsoids will interest us far more than the others as these are the ones which occur in most of the applications. For our purposes we need only consider cases where the centre of the curve or the surface is at the origin, and we can manage to work interpreting vector symbols **x**, **y**, etc. as denoting bound vectors based on the origin. This associates each point P in the space with its position vector. These vectors are, of course, added by the parallelogram rule, that is to say the sum of two vectors is simply the diagonal (passing through the origin) of the parallelogram which they define.

Elementary texts frequently chop and change between bound and free vectors without any very satisfactory explanation of the relationship between the two. Bound vectors and free vectors are actually two distinct, but isomorphic, systems. A theoretical discussion of this relationship may be found in Reference [1]. Because of this isomorphism they are easily confused. A few books, very correctly, insist that free vectors are to be regarded as translations of the space, as distinct from line segments in it, but no elementary text, to the writer's knowledge, adopts a notation and terminology which keep this distinction clear and interrelate the two ideas consistently.

Our restriction on the use of geometrical vectors to vectors bound at the origin, which is the most simple and obvious geometrical interpretation of algebraic n-ples of numbers, will occasionally necessitate the use of phrases such as 'the line through P parallel to **x**'—but this is a small price to pay for consistency.

In two dimensions any conic section with centre at the origin has an equation of the form

$$ax_1^2 + 2hx_1x_2 + bx_2^2 = \text{constant}.$$

The left-hand side may be written as a matrix product,

$$\begin{bmatrix} x_1 & x_2 \end{bmatrix} \begin{bmatrix} a & h \\ h & b \end{bmatrix} \begin{bmatrix} x_1 \\ x_2 \end{bmatrix}.$$

By working this way each central conic is associated with a symmetric matrix and vice versa.

By taking different constants on the right-hand side of the equation we get a family of conics which are *similar* and *similarly situated*, that is to say they are of the same shape, with their axes parallel (Fig. 2.6). This is easily seen by carrying out the transformation $\mathbf{x} \to k\mathbf{x}$. If **x** satisfies the equation

$$\mathbf{x}'\mathbf{A}\mathbf{x} = c_1$$

then $k\mathbf{x}$ satisfies

$$(k\mathbf{x})'\mathbf{A}(k\mathbf{x}) = c_2,$$

with $c_2 = k^2 c_1$, (verify). But the transformation $x \to k\mathbf{x}$ is simply an enlargement about the origin, every point moving out radially until it is k times its original distance away. For this reason, given a symmetric matrix $\mathbf{A}$ it is often convenient to picture the family of curves $\mathbf{x}'\mathbf{A}\mathbf{x}$ = constant, rather than just one of them, such as $\mathbf{x}'\mathbf{A}\mathbf{x} = 1$.

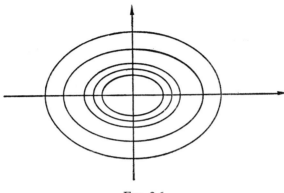

FIG. 2.6

Elementary books on geometry prove that the curve

$$ax_1^2 + 2hx_1x_2 + bx_2^2 = \text{constant}$$

is an ellipse if $h^2 < ab$ and a hyperbola if $h^2 > ab$. We will assume that this is known.

We will now see how to get information about tangents to a central conic whose equation is known, and then move on to consider tangent planes to surfaces.

Consider the ellipse with equation

$$x_1^2 + 6x_1x_2 + 4x_2^2 = 1.$$

In matrix notation this is written

$$[x_1 \ x_2] \begin{bmatrix} 1 & 3 \\ 3 & 4 \end{bmatrix} \begin{bmatrix} x_1 \\ x_2 \end{bmatrix} = 1. \qquad (2.25)$$

Differentiating in the usual way, with respect to x_1,

$$2x_1 + 6x_1\frac{dx_2}{dx_1} + 6x_2 + 8x_2\frac{dx_2}{dx_1} = 0.$$

From this we derive the gradient of the tangent,

$$\frac{dx_2}{dx_1} = -\frac{2x_1 + 6x_2}{6x_1 + 8x_2} = -\frac{x_1 + 3x_2}{3x_1 + 4x_2}.$$

Hence the gradient of the normal at the point (p, q) (Fig. 2.7) is equal to $(3p + 4q)/(p + 3q)$. This means that if a given segment of the

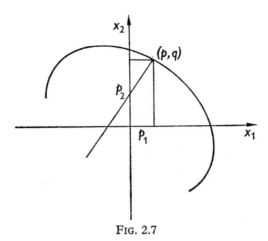

FIG. 2.7

normal at (p, q) is projected on to the x_1 and x_2 axes then the lengths of the projections are in the ratio

$$p_1 : p_2 = (p + 3q) : (3p + 4q).$$

The expressions in this ratio are given by the matrix product

$$\begin{bmatrix} 1 & 3 \\ 3 & 4 \end{bmatrix} \begin{bmatrix} p \\ q \end{bmatrix}; \qquad (2.26)$$

and so the matrix $\begin{bmatrix} 1 & 3 \\ 3 & 4 \end{bmatrix}$ is seen to have another connexion with the curve. In Eq. (2.25) the matrix specifies the curve in a certain way. In the expression (2.26) the matrix is seen as transforming the components of the position vector of the point (p, q) into the direction ratios of the normal. Putting it more briefly, the matrix maps the position vector on to the normal. If ξ is the position vector of a point on the curve then $A\xi$ is in the direction of the normal at the point (Fig. 2.8). Note that

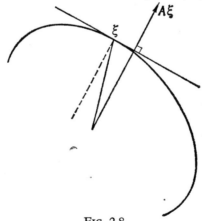

FIG. 2.8

nothing is being said about the actual lengths of the vectors ξ and $A\xi$ —we are only concerned with directions.

At first sight this way of looking at things may seem more complicated than the familiar methods of using gradients; but the present point of view extends easily to any number of dimensions. To show this we will now give a general proof of the property. It is convenient to think in terms of three dimensions but the following proof is valid in any number of dimensions. This means only that the algebra is valid however many components the vectors may have.

Our concern is with quadric surfaces with centre at the origin. By definition such a surface is the set of points **x** satisfying the relation

$$\mathbf{x'Ax} = \text{constant}, k \text{ (say)},$$

where **A** is a symmetric matrix. In two-dimensional space the loci are ellipses (including circles) and hyperbolas (including pairs of lines when $k = 0$). In three dimensions the loci are ellipsoids (by far the most important as far as we are concerned), the two kinds of hyperboloid, and various kinds of cone (when $k = 0$). Special names will not be necessary in spaces of higher dimension; it will be enough to speak by analogy with the three dimensional case. When referring to 'quadrics' later in this chapter 'conics' are usually included as well, but it is a convenience not to employ roundabout phraseology to say this every time. Also as our concern is exclusively with central quadrics it is convenient to write briefly of 'quadrics' meaning 'central quadrics' only. Any departure from this usage should be clear from the context.

THEOREM 2.2

If ξ is a point on the surface $\mathbf{x'Ax} = k$ (where $\mathbf{A}$ is a symmetric matrix), then the normal at ξ is in the direction of $\mathbf{A}\xi$.

Proof

Let ξ be the position vector of a point P on the surface and η the position vector of any other point (Fig. 2.9). The idea underlying the proof is to find the condition for a line through ξ parallel to η to be a tangent to the surface.

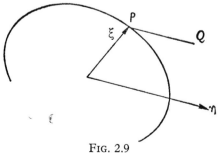

Fig. 2.9

For the moment let ξ be any point, on the surface or not. Then any other point Q on the line through P parallel to η has position vector $\xi + t\eta$, where t is a scalar parameter. If this is on the surface then

$$(\xi + t\eta)'\mathbf{A}(\xi + t\eta) = k,$$
$$(\xi' + t\eta')\mathbf{A}(\xi + t\eta) = k,$$

using the easily proved result that $(\xi + t\eta)' = \xi' + t\eta'$.

Expanding out, and collecting the terms together according to the powers of t,

$$\xi'\mathbf{A}\xi + t(\eta'\mathbf{A}\xi + \xi'\mathbf{A}\eta) + t^2\eta'\mathbf{A}\eta = k.$$

This is a quadratic equation for t, giving the two points where the line PQ cuts the surface. If ξ is on the surface $\xi'\mathbf{A}\xi = k$, the equation reduces to

$$t(\eta'\mathbf{A}\xi + \xi'\mathbf{A}\eta) + t^2\eta'\mathbf{A}\eta = 0,$$

and one value of t is zero. For the line to be a tangent the other value of t must be zero also, and this requires

$$\eta'\mathbf{A}\xi + \xi'\mathbf{A}\eta = 0.$$

At this stage of the argument the symmetry of $\mathbf{A}$ is involved. $\xi'\mathbf{A}\eta$ is a 1×1 matrix and so is equal to its own transpose. Applying a result in Exercise 2.13 and using the symmetry of $\mathbf{A}$

$$\xi'\mathbf{A}\eta = (\xi'\mathbf{A}\eta)' = \eta'\mathbf{A}'\xi = \eta'\mathbf{A}\xi;$$

that is to say the two terms on the left-hand side of the above equation are equal. This means that

$$\eta'\mathbf{A}\xi = \xi'\mathbf{A}\eta = 0;$$

and this implies that η is perpendicular to $\mathbf{A}\xi$ and also that ξ is perpendicular to $\mathbf{A}\eta$. (We are using the result that two vectors are perpendicular if and only if their inner product vanishes.)

In particular, we see that if η is given the direction of a tangent through P then it is perpendicular to $\mathbf{A}\xi$. But only one line is perpendicular to the set of tangents through P, and that is the normal at P.

Hence the normal at the point with position vector ξ has direction $\mathbf{A}\xi$.

Exercise 2.25

Draw various central conics in two dimensions, and for various vectors ξ draw the corresponding vector $\mathbf{A}\xi$. Verify the theorem which has just been proved.

2.8 Principal axes and orthogonal transformations

The axes of symmetry are a prominent geometrical feature of central conics in two dimensions and of central quadrics in three. When more dimensions are involved the axes are defined algebraically, by methods which apply in any number of dimensions. A particular case will show the general method of finding the axes of symmetry of a central quadric from its algebraic equation. An example based on an ellipsoid is taken because ellipsoids occur in the applications far more frequently than hyperboloids.

EXAMPLE 2.13

Consider the quadric surface

$$6x_1^2 + 8x_1x_2 - 4x_1x_3 + 12x_2^2 - 8x_2x_3 + 13x_3^2 = k.$$

In matrix notation this is

$$\mathbf{x'Ax} = k \quad \text{with} \quad \mathbf{A} = \begin{bmatrix} 6 & 4 & -2 \\ 4 & 12 & -4 \\ -2 & -4 & 13 \end{bmatrix}.$$

We know that the normal at $\mathbf{x}$ has direction $\mathbf{Ax}$. What distinguishes the points of the surface which are on the axes of symmetry? At these the normal lies along the position vector. This means that the axes of symmetry lie along directions $\mathbf{x}$, for which $\mathbf{Ax}$ is a scalar multiple of $\mathbf{x}$.

This immediately raises the question discussed in Section 2.5. Given a matrix $\mathbf{A}$, can we find a scalar λ and a vector $\mathbf{x}$ such that $\mathbf{Ax} = \lambda\mathbf{x}$? The axes of symmetry of the quadric are simply the eigenvectors of the associated matrix.

If $\mathbf{Ax} = \lambda\mathbf{x}$, then in this case

$$\begin{aligned} 6x_1 + 4x_2 - 2x_3 &= \lambda x_1, \\ 4x_1 + 12x_2 - 4x_3 &= \lambda x_2, \\ -2x_1 - 4x_2 + 13x_3 &= \lambda x_3. \end{aligned}$$

Hence

$$\begin{aligned} (6-\lambda)x_1 + 4x_2 - 2x_3 &= 0, \\ 4x_1 + (12-\lambda)x_2 - 4x_3 &= 0, \\ -2x_1 - 4x_2 + (13-\lambda)x_3 &= 0. \end{aligned} \quad (2.27)$$

Using the previously quoted result on determinants,

$$\begin{vmatrix} 6-\lambda & 4 & -2 \\ 4 & 12-\lambda & -4 \\ -2 & -4 & 13-\lambda \end{vmatrix} = 0.$$

Tedious manipulation, of which we do not give the details, leads to the three values of λ which satisfy this cubic equation. They are

$$\lambda = 4, 9, 18.$$

When $\lambda = 4$, Eqs. (2.27) are

$$2x_1 + 4x_2 - 2x_3 = 0,$$
$$4x_1 + 8x_2 - 4x_3 = 0,$$
$$-2x_1 - 4x_2 + 9x_3 = 0.$$

From these equations

$$x_1 : x_2 : x_3 = 2 : -1 : 0.$$

As always the eigenvectors are determined only as far as ratios, and not in magnitude. The eigenvalues are, of course, very definitely determined in magnitude, and it will transpire later that these magnitudes have a geometrical significance.

Likewise, when $\lambda = 9$, Eqs. (2.27) become

$$-3x_1 + 4x_2 - 2x_3 = 0,$$
$$4x_1 + 3x_2 - 4x_3 = 0,$$
$$-2x_1 - 4x_2 + 4x_3 = 0;$$

which have solution

$$x_1 : x_2 : x_3 = 2 : 4 : 5.$$

When $\lambda = 18$

$$x_1 : x_2 : x_3 = 1 : 2 : -2.$$

(Verify this.)

It is important to remember the significance of the eigenvectors in this context—they give the directions of the principal axes (axes of symmetry).

It is a familiar piece of geometry that the principal axes of a quadric surface are mutually orthogonal. This is easily verified in this case. Two vectors are orthogonal if their inner product is zero. For the first two eigenvectors we have

$$(2, -1, 0)(2, 4, 5)' = 0,$$

and so these are orthogonal. Similarly for the other two pairs.

In the general case it is easy to establish the following theorem.

THEOREM 2.3

If $\mathbf{A}$ is a symmetric matrix, and λ and μ are two distinct eigenvalues and $\mathbf{x}$ and $\mathbf{y}$ (respectively) are corresponding eigenvectors, then $\mathbf{x}$ and $\mathbf{y}$ are orthogonal.

Proof

From the definition of eigenvalue and eigenvector

$$\mathbf{A}\mathbf{x} = \lambda \mathbf{x} \quad \text{and} \quad \mathbf{A}\mathbf{y} = \mu \mathbf{y}.$$

Transposing,

$$\mathbf{y}'\mathbf{A}' = \mu \mathbf{y}'$$

and since $\mathbf{A}$ is symmetric

$$\mathbf{y}'\mathbf{A} = \mu \mathbf{y}'.$$

Post-multiplying by $\mathbf{x}$

$$\mathbf{y}'\mathbf{A}\mathbf{x} = \mu \mathbf{y}'\mathbf{x}.$$

But pre-multiplying $\mathbf{A}\mathbf{x} = \lambda \mathbf{x}$ by $\mathbf{y}'$, we get

$$\mathbf{y}'\mathbf{A}\mathbf{x} = \lambda \mathbf{y}'\mathbf{x}.$$

Hence

$$\lambda \mathbf{y}'\mathbf{x} = \mu \mathbf{y}'\mathbf{x}.$$

But $\lambda \neq \mu$, so

$$\mathbf{y}'\mathbf{x} = 0.$$

That is to say, $\mathbf{y}$ and $\mathbf{x}$ are orthogonal.

Observe that if $\mathbf{A}$ has a repeated eigenvalue there is a complication. This will be considered later.

Continuing with Example 2.13, the principal axes are an obvious choice as co-ordinate axes, since the equation of the surface then takes a particularly simple form. So what is the equation of the surface with respect to these axes? It may help some readers if we stress the point that we are now considering a change of co-ordinates and *not*, as in many of the previous examples, a point transformation, i.e. a transformation of the figure against the background of a fixed co-ordinate system.

The direction ratios of the first eigenvector are $(2, -1, 0)$, and if these are normalized to give direction cosines they become $(2/\sqrt{5}, -1/\sqrt{5}, 0)$. Likewise the direction cosines of the second and third eigenvectors are $(2/\sqrt{45}, 4/\sqrt{45}, 5/\sqrt{45})$ and $(1/3, 2/3, -2/3)$.

If the directions of the eigenvectors are taken as axes y_1, y_2 and y_3, then the y_1 co-ordinate of any point P is equal to the sum of the projections of its x co-ordinates in the direction y_1 (Fig. 2.10). This is because the projection of OP is equal to the sum of the projections of the segments OL, LM, MP. So the required y_1 co-ordinate is given by multiplying the co-ordinates x_1, x_2 and x_3 respectively by the appropriate direction cosines and adding. Similarly for the y_2 and y_3 co-ordinates.

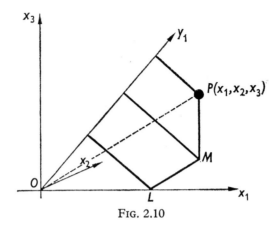

FIG. 2.10

It is convenient to tabulate the cosines of the angles between the y_1, y_2, y_3 axes and the x_1, x_2, x_3 axes as follows

	x_1	x_2	x_3
y_1	$2/\sqrt{5}$	$-1/\sqrt{5}$	0
y_2	$2/\sqrt{45}$	$4/\sqrt{45}$	$5/\sqrt{45}$
y_3	$1/3$	$2/3$	$-2/3$

Then
$$y_1 = \frac{2}{\sqrt{5}}x_1 - \frac{1}{\sqrt{5}}x_2.$$

Likewise
$$y_2 = \frac{2}{\sqrt{45}}x_1 + \frac{4}{\sqrt{45}}x_2 + \frac{5}{\sqrt{45}}x_3,$$

and
$$y_3 = \frac{1}{3}x_1 + \frac{2}{3}x_2 - \frac{2}{3}x_3.$$

In matrix form

$$\mathbf{y} = \begin{bmatrix} \dfrac{2}{\sqrt{5}} & -\dfrac{1}{\sqrt{5}} & 0 \\ \dfrac{2}{\sqrt{45}} & \dfrac{4}{\sqrt{45}} & \dfrac{5}{\sqrt{45}} \\ \dfrac{1}{3} & \dfrac{2}{3} & -\dfrac{2}{3} \end{bmatrix} \mathbf{x}.$$

This gives the new co-ordinates in terms of the old, but to transform the equation of the quadric to the new co-ordinate system it is actually easier to use the inverse transformation. This may be written out using the direction cosines of the x-axes with respect to the y-axes. These are also given by the previous table, using the columns instead of the rows. The inverse transformation is therefore

$$\mathbf{x} = \begin{bmatrix} \dfrac{2}{\sqrt{5}} & \dfrac{2}{\sqrt{45}} & \dfrac{1}{3} \\ -\dfrac{1}{\sqrt{5}} & \dfrac{4}{\sqrt{45}} & \dfrac{2}{3} \\ 0 & \dfrac{5}{\sqrt{45}} & -\dfrac{2}{3} \end{bmatrix} \mathbf{y},$$

or more fully,

$$x_1 = \frac{2}{\sqrt{5}} y_1 + \frac{2}{\sqrt{45}} y_2 + \frac{1}{3} y_3,$$

$$x_2 = -\frac{1}{\sqrt{5}} y_1 + \frac{4}{\sqrt{45}} y_2 + \frac{2}{3} y_3,$$

$$x_3 = \frac{5}{\sqrt{45}} y_2 - \frac{2}{3} y_3.$$

Verify that this is the inverse of the previous transformation.

When this co-ordinate change is made to the quadric surface

$$\mathbf{x}'\mathbf{A}\mathbf{x} = 6x_1^2 + 8x_1x_2 - 4x_1x_3 + 12x_2^2 - 8x_2x_3 + 13x_3^2 = k \quad (2.28)$$

it reduces to

$$4y_1^2 + 9y_2^2 + 18y_3^2 = k. \quad (2.29)$$

The verification of this is tedious, but it is very instructive to do one or two similar calculations at length, although this is clearly not a thing which one would wish to do every time. The purpose of matrix algebra is to provide general methods which avoid this kind of tedious detail, and a full appreciation of the gain to be had is achieved by doing some calculations the long, hard way. A general proof that the use of the eigenvectors in this way reduces the equation of the quadric surface to the sum of squares will be given shortly. The important thing is that the coefficients of the terms on the left-hand side of the equation are simply the eigenvalues of the matrix. If we know this is going to happen we can write down the expression immediately, and much of the above working will be unnecessary. It is not even necessary to calculate the eigenvectors if all that is required is the expression

$$4y_1^2 + 9y_2^2 + 18y_3^2 = k.$$

For positive values of k this is the equation of an ellipsoid. In particular, if $k = 1$, the lengths of the principal semi-axes are easily seen to be $1/\sqrt{4}$, $1/\sqrt{9}$ and $1/\sqrt{18}$. This is the significance of the eigenvalues of the matrix, they are related to the lengths of the principal axes of the quadric. In general, if a quadric surface in n dimensions has equation $\mathbf{x}'\mathbf{A}\mathbf{x} = 1$, then the lengths of the semi-axes are given by $1/\sqrt{\lambda_1}, \ldots, 1/\sqrt{\lambda_n}$ where $\lambda_1, \ldots, \lambda_n$ are the eigenvalues of $\mathbf{A}$. Observe that a *large* eigenvalue corresponds to a *short* axis. If $\mathbf{A}$ has a zero eigenvalue there are obvious complications. These exceptional cases will have to be considered on their merits as they arise.

The matrix of direction cosines has some special properties. Call it $\mathbf{H}$. Then we have seen that the transformation $\mathbf{y} = \mathbf{H}\mathbf{x}$ has inverse $\mathbf{x} = \mathbf{H}'\mathbf{y}$. In other words, the inverse of this matrix is the same as its transpose. Matrices with this property are called *orthogonal* matrices. The name comes from the present application. It may be noted that the word *orthogonal* is used in different ways in different contexts. The word has one meaning when applied to vectors and another when applied to matrices. Bearing in mind that the calculation of an inverse matrix is usually a lengthy and awkward process, it can be seen that orthogonal matrices are unusually convenient.

To summarize, an orthogonal matrix $\mathbf{H}$ is a matrix with the special property
$$\mathbf{H}'\mathbf{H} = \mathbf{H}\mathbf{H}' = \mathbf{I}.$$

Exercise 2.26

Prove that the product of two orthogonal matrices is orthogonal.

It is now necessary to consider the transformation of Eq. (2.28) into Eq. (2.29) more generally. This transformation has two closely related aspects. The quadratic form on the left of Eq. (2.28) is transformed into the quadratic form on the left of Eq. (2.29); but this is equivalent to the transformation of the matrix

$$\mathbf{A} = \begin{bmatrix} 6 & 4 & -2 \\ 4 & 12 & -4 \\ -2 & -4 & 13 \end{bmatrix}$$

into the matrix

$$\mathbf{D} = \begin{bmatrix} 4 & 0 & 0 \\ 0 & 9 & 0 \\ 0 & 0 & 18 \end{bmatrix}.$$

The quadratic expression $\mathbf{x}'\mathbf{A}\mathbf{x}$ becomes $\mathbf{y}'\mathbf{D}\mathbf{y}$ on making the substitution $\mathbf{x} = \mathbf{H}\mathbf{y}$. But putting $\mathbf{x} = \mathbf{H}\mathbf{y}$ into $\mathbf{x}'\mathbf{A}\mathbf{x}$ gives

$$(\mathbf{y}'\mathbf{H}')\mathbf{A}(\mathbf{H}\mathbf{y}) = \mathbf{y}'\mathbf{H}'\mathbf{A}\mathbf{H}\mathbf{y};$$

and so the problem, in general terms, is to show that given the symmetric matrix **A**, the orthogonal matrix **H**, constructed as it was, has the property that **H'AH** is a diagonal matrix, of which the diagonal elements are equal to the eigenvalues of **A**. The general algebraic proof of this property turns out to be much easier than the arithmetic in the earlier, typical example.

THEOREM 2.4

*Given a symmetric matrix **A** and the orthogonal matrix **H** composed of the normalized eigenvectors of **A**, then the matrix **H'AH** is a diagonal matrix which has as diagonal elements the eigenvalues of **A**.*

The proof which follows will assume that **A** has distinct eigenvalues.

Proof

For convenience we will write the proof out for a 3×3 matrix; a similar method applies to a square matrix of any size.

Let the eigenvalues of **A** be λ, μ and ν, and the corresponding eigenvectors ξ, η and ζ.

Put $$\mathbf{H} = [\xi \mid \eta \mid \zeta].$$

This notation means that **H** is built out of the column vectors ξ, η and ζ.

Then $$\mathbf{AH} = \mathbf{A}[\xi \mid \eta \mid \zeta] = [\mathbf{A}\xi \mid \mathbf{A}\eta \mid \mathbf{A}\zeta].$$

This is a result of the definition of matrix multiplication as multiplication by row and column. The reader who is in doubt is advised to rewrite this whole equation in full, using 3×3 matrices.

But by the eigenvector property the right-hand side reduces, giving

$$\mathbf{AH} = [\lambda\xi \mid \mu\eta \mid \nu\zeta].$$

The right-hand side can be rewritten once more, as

$$\mathbf{AH} = [\xi \mid \eta \mid \zeta]\begin{bmatrix} \lambda & 0 & 0 \\ 0 & \mu & 0 \\ 0 & 0 & \nu \end{bmatrix}.$$

Again, if in doubt write all the expressions out in full as 3×3 matrices. This reduces at once to

$$\mathbf{AH} = \mathbf{HD}.$$

where **D** is the diagonal matrix with λ, μ, ν along the principal diagonal. Now, using the property that $\mathbf{H'H} = \mathbf{I}$, we have

$$\mathbf{H'AH} = \mathbf{D}.$$

This reduction of **A** to diagonal form is a piece of algebra with many applications, and it may be well to pause at this point to make a number of observations.

The above method of proof does assume that the eigenvalues are distinct, because Theorem 2.3 has been used when constructing the orthogonal matrix **H**. If **A** has distinct eigenvalues then Theorem 2.3 ensures that there are n (in the general case) mutually orthogonal eigenvectors with which to construct **H**. It may still be possible to construct a suitable **H** even if some eigenvalues of **A** are repeated. This is shown in the example which follows shortly. A general algebraic proof that the procedure of this example is always possible and always valid is more difficult. For this more advanced books must be consulted, for example that by Mirsky [4].

The following related theorem is also involved in later developments.

THEOREM 2.5

*If **A** is a symmetric matrix and **H** is any orthogonal matrix then **H'AH** is also symmetric and **H'AH** and **A** have the same eigenvalues.*

Proof

Let λ be an eigenvalue of **A** and **x** a corresponding eigenvector. Then, putting $\mathbf{x} = \mathbf{Hy}$

$$\mathbf{Ax} = \lambda \mathbf{x} \Rightarrow \mathbf{AHy} = \lambda \mathbf{Hy},$$

$$\Rightarrow \mathbf{H'AHy} = \lambda \mathbf{y}, \quad \text{since} \quad \mathbf{H}^{-1} = \mathbf{H'}.$$

This shows that $\mathbf{y} = \mathbf{Hx}$ is an eigenvector of **H'AH**, with eigenvalue λ. It is necessary, however, to ensure that $\mathbf{y} \neq 0$, and this may be proved as follows.

$$\mathbf{y} = 0 \Rightarrow \mathbf{H'y} = \mathbf{H'Hx} = \mathbf{Ix} = \mathbf{x} = 0.$$

But by hypothesis $\mathbf{x} \neq 0$. Hence any eigenvalue of **A** is also an eigenvalue of **H'AH**. A similar proof shows that any eigenvalue of **H'AH** is also an eigenvalue of **A**.

Much of the above work can be extended to cases where **A** is not symmetric. This does not concern us here, but it is resumed in Section 6.6.

EXAMPLE 2.14

Consider the matrix in Exercise 2.22:

$$\mathbf{A} = \begin{bmatrix} 5 & 1 & 1 \\ 1 & 5 & -1 \\ 1 & -1 & 5 \end{bmatrix}$$

was shown to have an eigenvalue 3 with eigenvector $(-1, 1, 1)'$ and a double eigenvalue 6. A short search among the multiplicity of eigenvectors associated with the double eigenvalue will reveal a pair which are mutually orthogonal. $(1, 1, 0)'$ and $(1, -1, 2)'$ is such a pair, but there is an infinite choice of other pairs which would do just as well.

H can now be constructed as

$$\mathbf{H} = \begin{bmatrix} -\dfrac{1}{\sqrt{3}} & \dfrac{1}{\sqrt{2}} & \dfrac{1}{\sqrt{6}} \\ \dfrac{1}{\sqrt{3}} & \dfrac{1}{\sqrt{2}} & -\dfrac{1}{\sqrt{6}} \\ \dfrac{1}{\sqrt{3}} & 0 & \dfrac{2}{\sqrt{6}} \end{bmatrix}.$$

Verify now that

$$\mathbf{H'AH} = \mathrm{diag}\{3, 6, 6\}.$$

The notation $\mathrm{diag}\{3, 6, 6\}$ denotes the matrix in which the only non-zero elements are in the principal diagonal, and are 3, 6, 6.

In terms of the related quadric surface we see that the transformation $\mathbf{x} = \mathbf{Hy}$ transforms the equation

$$5x_1^2 + 2x_1x_2 + 2x_1x_3 + 5x_2^2 - 2x_2x_3 + 5x_3^2 = k$$

in the (x_1, x_2, x_3) co-ordinate system to the equation

$$3y_1^2 + 6y_2^2 + 6y_3^2 = k$$

in the (y_1, y_2, y_3) system.

This surface is a spheroid. In fact it is a prolate (i.e. elongated) spheroid. (Why?) The multiplicity of eigenvectors connected with the eigenvalue 6 is the algebraic counterpart of the multitude of transverse principal axes which this surface, rather like a Rugby football, has orthogonal to its main longitudinal axis. It may be as well to re-emphasize that the character of the surface can be decided from a knowledge of the eigenvalues of the matrix alone—no further calculation is necessary.

Exercise 2.27

Find the eigenvalues of the matrix

$$\mathbf{A} = \begin{bmatrix} 2 & -2 & 0 \\ -2 & 1 & -2 \\ 0 & -2 & 0 \end{bmatrix},$$

and find a matrix such that $\mathbf{H'AH}$ is diagonal. What shape are the related quadric surfaces?

Exercise 2.28
Answer the same questions for the matrix
$$\mathbf{A} = \begin{bmatrix} 1 & 1 & -1 \\ 1 & 0 & -2 \\ -1 & -2 & 0 \end{bmatrix}.$$

Exercise 2.29
Answer the same questions for the matrix
$$\mathbf{A} = \begin{bmatrix} 1 & 2 & 4 \\ 2 & -2 & 2 \\ 4 & 2 & 1 \end{bmatrix}.$$

Exercise 2.30
Verify that the matrix
$$\begin{bmatrix} 3 & 1 & -1 & 1 \\ 1 & 3 & -1 & 1 \\ -1 & -1 & 3 & -1 \\ 1 & 1 & -1 & 3 \end{bmatrix}$$
has an eigenvalue 6 and a triple eigenvalue 2. Find a matrix $\mathbf{H}$ which will diagonalize it.

Exercise 2.31
So far we have only considered orthogonal matrices in connexion with co-ordinate transformations. The importance of orthogonal matrices in connexion with point transformations is that they describe rigid body displacements about the origin (i.e. rotations about some axis through the origin or reflections in planes through the origin). This property follows very easily once the two following properties have been demonstrated:

(i) the transformation $\mathbf{x} \to \mathbf{Hx}$ leaves the length of the vector $\mathbf{x}$ unaltered, i.e. $\mathbf{x'x} = \mathbf{x'H'Hx}$;
(ii) the transformation $\mathbf{x} \to \mathbf{Hx}$ leaves the angle between two vectors unaltered.

Prove these properties.

An important, though little discussed, branch of pedagogy is the construction of numerical examples which work out easily. Have you ever wondered how it is that the textbook examples on eigenvalues work out with integer solutions, when the probability that this will happen when matrices are written down at random is extremely small? In Theorem 2.4 we had $\mathbf{H'AH} = \mathbf{D}$. But since $\mathbf{H'} = \mathbf{H}^{-1}$ we quickly derive $\mathbf{A} = \mathbf{HDH'}$.

This gives the following procedure for constructing a matrix with prescribed eigenvalues. Arrange the prescribed eigenvalues in diagonal matrix **D**. Take any set of mutually orthogonal vectors, normalize them and use them as the columns of an orthogonal matrix **H**. Then **HDH'** is a matrix with the prescribed eigenvalues. Try to construct some 3×3 matrices with integral eigenvalues, suitable for use as exercise examples.

2.9 Some further examples on mappings

The purpose of this section is to relate the work done on mappings $\mathbf{x} \to \mathbf{Mx}$ with the work on quadrics $\mathbf{x'Ax} = k$. It transpires that important connections arise when $\mathbf{A} = \mathbf{M'M}$.

When certain comparatively simple properties of loci of the form $\mathbf{x'Ax} = k$ have been understood in two and three dimensions, then further properties quickly follow, being immediately obvious from a geometrical point of view. An algebraic proof of the corresponding algebraic relations is sometimes not quite so easy, and in these cases a geometrical understanding of what is involved is a valuable supplementary insight. We will now consider two problems which can occur in a purely algebraic form in a number of applications (for example in statistics, as in Section 7.6), but we will consider them from a geometrical point of view.

In the paragraphs which follow it simplifies the matter to assume, initially at least, that the eigenvalues of **A** are positive and distinct. This is usually the case in many practical applications. The venturesome reader may seek to remove these restrictions if he wishes.

Problem 1

Consider the locus $\mathbf{x'Ax} = k$, $(\mathbf{A'} = \mathbf{A})$. Where on the locus is $\mathbf{x'x}$ a maximum?

When the eigenvalues are all positive the locus is an ellipsoid, and the question asks simply which point on the locus is the furthest from the origin; and it is geometrically obvious that the answer is provided by the two points at opposite ends of the major axis. This axis is, you will remember, in the direction of the eigenvector of **A** corresponding to $\lambda_{\min}$.

We may then ask a supplementary question. If we consider the points on the locus with position vector **y** orthogonal to the vector **x** just found, where is $\mathbf{y'y}$ a maximum? The answer this time is provided by the points at the extremities of the next axis, in decreasing order of length. By imposing an extra restriction at each stage this maximizing question singles out the various axes of the surface in turn, in an order which corresponds to the eigenvalues in *increasing* order of magnitude.

Similar questions could be asked about minima, starting with the shortest axis.

Problem 2

Consider the locus $\mathbf{x}'\mathbf{x} = 1$, which is a circle in two dimensions, and we will call it a 'sphere' in more. Given a symmetric matrix $\mathbf{A}$, where on $\mathbf{x}'\mathbf{x} = 1$ is $\mathbf{x}'\mathbf{A}\mathbf{x}$ a maximum?

From a geometrical point of view we need to consider the surfaces $\mathbf{x}'\mathbf{A}\mathbf{x} = k$ in relation to the sphere $\mathbf{x}'\mathbf{x} = 1$. For varying values of k $\mathbf{x}'\mathbf{A}\mathbf{x} = k$ gives a family of ellipsoids. All of these ellipsoids have the same principal axes, but the larger the value of k the larger the ellipsoid.

For all sufficiently large values of k the ellipsoid is completely outside the sphere, but if we reduce k eventually the ellipsoid just touches the sphere, and it touches at the end of its shortest axis of symmetry (Fig. 2.11), the axis corresponding to the largest eigenvalue of $\mathbf{A}$. Therefore

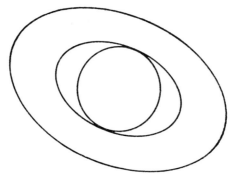

Fig. 2.11

the answer to the question formulated as Problem 2 is that $\mathbf{x}'\mathbf{A}\mathbf{x}$ is a maximum at the two diametrically opposite points on the sphere located by the eigenvector corresponding to the largest eigenvalue of $\mathbf{A}$.

We may follow with a supplementary question. Among the points on the sphere with position vector orthogonal to the vector $\mathbf{x}$ just found, where is $\mathbf{y}'\mathbf{A}\mathbf{y}$ a maximum? Geometrical considerations show that the answer is provided by the diametrically opposite points located by the eigenvector corresponding to the second biggest eigenvalue of $\mathbf{A}$.

Proceeding in this way the various eigenvalues and eigenvectors can be obtained successively as the solutions to a series of maximization problems. Equally well we could have asked questions about minima, and identified the eigenvectors in reverse order.

In particular $\mathbf{x}'\mathbf{A}\mathbf{x} = \lambda_{max}$ touches $\mathbf{x}'\mathbf{x} = 1$ externally at the point on $\mathbf{x}'\mathbf{x} = 1$ where $\mathbf{x}'\mathbf{A}\mathbf{x}$ is a maximum. Likewise $\mathbf{x}'\mathbf{A}\mathbf{x} = \lambda_{min}$ touches it internally where $\mathbf{x}'\mathbf{A}\mathbf{x}$ is a minimum. For intermediate eigenvalues similar results hold, but the contact is more complicated, being rather like that of the tangent plane at a saddle-point on a surface. Figure 2.12, showing the case in two dimensions, is the best that can be managed as an illustration.

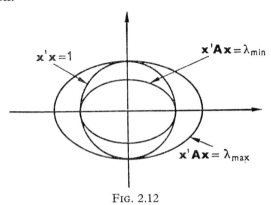

FIG. 2.12

The geometrical pictures are very valuable, but it is also necessary to consider an algebraic approach to these problems. Working algebraically it is just as easy to consider negative eigenvalues as it is positive, and so the restriction to positive eigenvalues will now be dropped. By considering the ratio $\mathbf{x'Ax}/\mathbf{x'x}$ Problems 1 and 2 can be treated at the same time.

Remembering Theorem 2.4 and the related work we know that given a symmetric matrix $\mathbf{A}$ with eigenvalues $\lambda_1, \lambda_2, \ldots, \lambda_n$ we may find an orthogonal matrix $\mathbf{H}$ such that $\mathbf{H'AH} = \operatorname{diag}\{\lambda_1, \lambda_2, \ldots, \lambda_n\}$. This was proved for the case when the eigenvalues are distinct; the result is true in general, but this was not proved although indications were given of how repeated eigenvalues can be handled.

Using the transformation $\mathbf{y} = \mathbf{Hx}$, which means going over to a co-ordinate system with the principal axes as co-ordinate axes, the quadratic form describing the related quadrics becomes

$$\mathbf{x'Ax} = \mathbf{y'H'AHy} = \mathbf{y'Dy}, \quad \text{where } \mathbf{D} = \operatorname{diag}\{\lambda_1, \lambda_2, \ldots, \lambda_n\},$$
$$= \lambda_1 y_1^2 + \lambda_2 y_2^2 + \ldots + \lambda_n y_n^2.$$

At the same time

$$\mathbf{x'x} = \mathbf{y'H'Hy} = \mathbf{y'y}, \quad \text{since } \mathbf{H'H} = \mathbf{I},$$
$$= y_1^2 + y_2^2 + \ldots + y_n^2.$$

This algebra enables us to establish the following theorem.

THEOREM 2.6

Given a symmetric matrix $\mathbf{A}$ the quotient $\mathbf{x'Ax}/\mathbf{x'x}$ has values which are between $\lambda_{\max}$ and $\lambda_{\min}$, the maximum and minimum eigenvalues of $\mathbf{A}$.

The quotient has the value $\lambda_{\max}$ when **x** *is an eigenvector corresponding to the eigenvalue $\lambda_{\max}$, and has the value $\lambda_{\min}$ when* **x** *is an eigenvector corresponding to the eigenvalue $\lambda_{\min}$.*

Proof

It is convenient to assume for the moment that the eigenvalues are distinct and label them so that $\lambda_1 > \lambda_2 > ... > \lambda_n$. Then putting

$$Q = \mathbf{x'Ax}/\mathbf{x'x} = \mathbf{y'H'AHy}/\mathbf{y'y},$$

$$\lambda_1 - Q = \frac{(\lambda_1 - \lambda_1)y_1^2 + (\lambda_1 - \lambda_2)y_2^2 + ... + (\lambda_1 - \lambda_n)y_n^2}{y_1^2 + y_2^2 + ... + y_n^2}.$$

The denominator is always positive when $\mathbf{y} \neq 0$. The coefficients in the numerator are all positive except the first which is zero, so that the numerator is greater than zero except when

$$y_2 = ... = y_n = 0.$$

Hence, always $Q \leqslant \lambda_1$, and equality occurs only when **y** is the vector $(y_1, 0, ..., 0)$, which in this co-ordinate system is the eigenvector with eigenvalue λ_1.

This proves the result for $\lambda_{\max}$ and a similar proof applies to $\lambda_{\min}$. Adaptations may be made to cover cases of repeated eigenvalues. There is no problem when intermediate eigenvalues are repeated, but when $\lambda_{\max}$ or $\lambda_{\min}$ is repeated a little care is required, and we will not give space to the details.

It is not difficult to go a little further and to show that in the neighbourhood of an eigenvector the quotient $\mathbf{x'Ax}/\mathbf{x'x}$, which is called the *Rayleigh quotient*, differs from the eigenvalue by a small quantity of the second order. This property is helpful when estimating eigenvalues because it means that comparatively coarse approximations in terms of eigenvectors give rather better approximations to eigenvalues than might have been expected. We do not intend to make any explicit application of this property and so we will not prove it. It may be observed, however, that this property and other similar properties of the Rayleigh quotient are almost obvious when seen in terms of the appropriate geometrical picture. The student is most likely to encounter the Rayleigh quotient in a mechanical context, and writers on mechanics seldom bring out the geometrical background. Writers on mechanics often refer to *Rayleigh's principle*. This may be enunciated in a variety of ways, but for our purposes we may regard Rayleigh's principle as being expressed by Theorem 2.6. Slight modifications have to be made later when dealing with systems with an infinite number of eigenvalues.

Exercise 2.32
Give an algebraic proof of a result which was stated earlier on a basis of geometrical intuition—that among all vectors orthogonal to the eigenvector corresponding to $\lambda_{\max}$, the vector which maximizes the quotient $\mathbf{x}'\mathbf{A}\mathbf{x}/\mathbf{x}'\mathbf{x}$ is the eigenvector corresponding to the next largest eigenvalue. (The method employed in the proof of Theorem 2.6 may be adapted.)

Exercise 2.33 (harder)
Consider the situation when some eigenvalues are zero, and consider the complications which arise with equal eigenvalues. How do the geometrical pictures apply when some eigenvalues are negative?

The importance of Problems 1 and 2 is that they provide a characterization of the eigenvalues and eigenvectors of **A** which is applicable in any number of dimensions and which does not depend on geometrical notions. Of course, the geometrical notions provide a most valuable mental picture; the point is that the argument does not depend on geometrical intuition. We may note also that whilst the argument is predominantly algebraic it involves other notions as well. The idea that a function attains its maximum value over a given domain is essentially a part of *analysis* and not a part of *algebra*. If eigenvalues are found, as they usually are in elementary exercises, by expanding the determinantal equation $|\mathbf{A} - \lambda\mathbf{I}| = 0$ and solving the resulting algebraic equation, the student may overlook that what is involved in principle is analysis; because only in very special cases can the roots of a polynomial equation be extracted by algebraic processes. The introduction of the ideas of eigenvalues into matrix algebra is a step beyond algebra into analysis.

Theorem 2.6 provides the basis for a definition of eigenvalue and eigenvector which is adaptable to a wider range of circumstance than some of the more elementary, and apparently simpler, definitions. It may be noted that it also provides, in principle, a way of computing the eigenvalues and eigenvectors of a symmetric matrix, but methods based on these ideas may require a great deal of computing power.

We now consider a number of examples which show mappings of various kinds. They provide empirical evidence which suggests the theorem which follows. These ideas are used in a statistical context in Section 7.3. Details of the working are frequently left to the reader. These examples may be omitted by readers who are not interested in the statistical applications.

R^2 denotes two-dimensional Cartesian space and R^3 denotes three-dimensional Cartesian space. The first few examples concern maps from R^2 to R^2, and we wish to find the image of the unit circle.

EXAMPLE 2.15

$$\mathbf{x} \to \mathbf{Mx}, \qquad \mathbf{M} = \begin{bmatrix} 1 & 0 \\ 0 & 0 \end{bmatrix}.$$

The unit circle is mapped on to the segment $(-1, 1)$ of the x_1-axis.

EXAMPLE 2.16

$$\mathbf{x} \to \mathbf{Mx}, \qquad \mathbf{M} = \begin{bmatrix} 0 & 1 \\ -1 & 0 \end{bmatrix}.$$

This is a rotation. The unit circle is rotated through a right angle.

EXAMPLE 2.17

$$\mathbf{x} \to \mathbf{Mx}, \qquad \mathbf{M} = \begin{bmatrix} 2 & 0 \\ 0 & 1 \end{bmatrix}.$$

This is a stretch. The unit circle becomes an ellipse with major axis 4, in the x_1 direction, and minor axis 2.

EXAMPLE 2.18

$$\mathbf{x} \to \mathbf{Mx}, \qquad \mathbf{M} = \begin{bmatrix} 1 & 1 \\ 0 & 1 \end{bmatrix}.$$

This is a shear. The unit circle becomes the ellipse

$$x^2 - 2x_1 x_2 + 2x_2^2 = 1.$$

The major axis is in the direction $(2, -1 + \sqrt{5})$ and is of length $1 + \sqrt{5}$, the minor axis is in the direction $(2, -1 - \sqrt{5})$ and is of length $-1 + \sqrt{5}$.

EXAMPLE 2.19

$$\mathbf{x} \to \mathbf{Mx}, \qquad \mathbf{M} = \begin{bmatrix} 0 & 1 \\ 0 & 1 \end{bmatrix}.$$

The unit circle is projected onto the line segment joining $(1, 1)$ to $(-1, -1)$. This has length $2\sqrt{2}$.

EXAMPLE 2.20
This concerns a map from R^2 into R^1.

$$\mathbf{x} \to \mathbf{Mx}, \qquad \mathbf{M} = [3 \quad 4].$$

The unit circle is mapped onto the line segment $(-5, 5)$. This is seen most easily by expressing points on the unit circle parametrically as $x_1 = \cos \theta$, $x_2 = \sin \theta$.

Then
$$3\cos\theta + 4\sin\theta = 5\cos(\theta - \alpha), \quad \text{where } \tan\alpha = \tfrac{4}{3}.$$
As θ varies, this function varies between $+5$ and -5.

EXAMPLE 2.21

This concerns a map from R^2 into R^3
$$\mathbf{x} \to \mathbf{Mx}, \quad \mathbf{M} = \begin{bmatrix} 1 & 0 \\ 0 & 1 \\ 0 & 1 \end{bmatrix}.$$

The point $x_1 = \cos\theta$, $x_2 = \sin\theta$ is mapped to $(\cos\theta, \sin\theta, \sin\theta)$ and it is not difficult to show that this is an ellipse in 3-space with axes $2\sqrt{2}$ and 2.

EXAMPLE 2.22

This concerns another map from R^2 into R^3
$$\mathbf{x} \to \mathbf{Mx}, \quad \mathbf{M} = \begin{bmatrix} 3 & 4 \\ 6 & 8 \\ 9 & 12 \end{bmatrix}.$$

The unit circle maps onto the line segment joining $(5, 10, 15)$ to $(-5, -10, -15)$.

The next two examples concern maps from R^3 onto R^3.

EXAMPLE 2.23
$$\mathbf{x} \to \mathbf{Mx}, \quad \mathbf{M} = \begin{bmatrix} 2 & 0 & 0 \\ 0 & 1 & 0 \\ 0 & 0 & 1 \end{bmatrix}.$$

This is a stretch. The unit sphere becomes a spheroid with axes 4, 2, 2.

EXAMPLE 2.24
$$\mathbf{x} \to \mathbf{Mx}, \quad \mathbf{M} = \begin{bmatrix} 1 & 0 & 0 \\ 0 & 1 & 0 \\ 0 & 0 & 0 \end{bmatrix}.$$

The unit sphere maps onto a circular disc, in the x_3 plane. The disc is the region $x_1^2 + x_2^2 \leqslant 1$.

The remaining examples are maps from R^3 into R^2.

EXAMPLE 2.25
$$\mathbf{x} \to \mathbf{Mx}, \quad \mathbf{M} = \begin{bmatrix} 1 & 0 & 0 \\ 0 & 1 & 0 \end{bmatrix}.$$

The unit sphere maps onto the unit disc (i.e. unit circle plus interior) in the image plane.

EXAMPLE 2.26

$$\mathbf{x} \to \mathbf{M}\mathbf{x}, \quad \mathbf{M} = \begin{bmatrix} 2 & 0 & 0 \\ 0 & 1 & 0 \end{bmatrix}.$$

The unit sphere maps onto an elliptic disc.

EXAMPLE 2.27

$$\mathbf{x} \to \mathbf{M}\mathbf{x}, \quad \mathbf{M} = \begin{bmatrix} 1 & 1 & 0 \\ 0 & 1 & 0 \end{bmatrix}.$$

The unit sphere is again mapped onto an elliptic disc.

EXAMPLE 2.28

$$\mathbf{x} \to \mathbf{M}\mathbf{x}, \quad \mathbf{M} = \begin{bmatrix} 2 & 2 & 1 \\ 4 & 4 & 2 \end{bmatrix}.$$

The unit sphere maps onto a line segment in the image plane.

The overriding problem is to generalize about the above examples. The initial vectors satisfy the restriction $\mathbf{x}'\mathbf{x} = 1$, but this relation cannot always be used to find an equation for the image locus. If we put $\mathbf{M}\mathbf{x} = \mathbf{y}$, then if $\mathbf{M}$ has an inverse we easily see that

$$\mathbf{x} = \mathbf{M}^{-1}\mathbf{y},$$

and so $\mathbf{y}$ satisfies the equation

$$\mathbf{y}'(\mathbf{M}^{-1})'\mathbf{M}^{-1}\mathbf{y} = 1,$$

which is a central conic in two dimensions or a central quadric in more dimensions. However, this argument does not apply to many of the above examples because in these $\mathbf{M}$ does not have an inverse. It can be seen that there are two sorts of reasons for this. In some cases the image is 'collapsed', forming typically a disc when we might have expected an ellipsoid; and in other cases it forms something like an elliptic line in a space which is 'big enough to hold something more'. These degeneracies make it hard to generalize.

It helps if we consider the maximum and minimum distances of the image locus from the origin in the image space. If $\mathbf{M}\mathbf{x} = \mathbf{y}$ we are asking the maximum and minimum values of $\mathbf{y}'\mathbf{y} = \mathbf{x}'\mathbf{M}'\mathbf{M}\mathbf{x}$, subject to the original constraint $\mathbf{x}'\mathbf{x} = 1$. In other words, we are led back to Problem 2, with $\mathbf{M}'\mathbf{M} = \mathbf{A}$.

It is easy to show that whatever $\mathbf{M}$ is (even if it is not square) $\mathbf{M}'\mathbf{M}$

is square and symmetric, and furthermore if the elements of **M** are all real then **M'M** has no negative eigenvalues.

Exercise 2.34
Prove this.

The solution to Problem 2 is therefore the key to the present situation. All of the above examples are covered by the following theorem.

THEOREM 2.7

The map $\mathbf{x} \to \mathbf{Mx} = \mathbf{y}$ *maps the unit sphere in x-space onto an ellipsoid in y-space, and the squares of the semi-axes of the image are equal to the eigenvalues of* **M'M**.

Proof
The details of the proof are left as an exercise.

Another question arises naturally in connexion with the map $\mathbf{x} \to \mathbf{Mx}$. We may ask about the locus in the original space which is mapped into the unit sphere in the object space. This question is closely connected with Problem 1, because if we have the map $\mathbf{x} \to \mathbf{Mx}$ and put $\mathbf{Mx} = \mathbf{y}$, then the **x**'s which are mapped into $\mathbf{y'y} = 1$ satisfy $\mathbf{x'M'Mx} = 1$, so they lie on a quadric surface in the original space. (Of course, this is a conic section in the two-dimensional case.) Note that the map may only be *into*, as distinct from *onto*, because not all points **y** may have any point **x** which corresponds.

Let us now reconsider Examples 2.15 to 2.28. We will investigate the nature of the locus $\mathbf{x'M'Mx} = 1$ and consider its image under the map. Does it cover the unit sphere in the image space or not? Observe that, in accordance with previous work the semi-axes of the locus are given by the reciprocals of the square roots of the eigenvalues of **M'M**, and note the limiting cases when **M'M** has zero eigenvalues. The details of the working are left as exercises.

EXAMPLE 2.15 (continued)
The locus $\mathbf{x'M'Mx} = 1$ in this case is the pair of lines $x_1 = \pm 1$. These map onto the two points $(\pm 1, 0)$.
M'M has eigenvalues 1 and 0.

EXAMPLE 2.16 (continued)
The unit circle is invariant. $\mathbf{M'M} = \mathbf{I}$.

EXAMPLE 2.17 (continued)
The ellipse $4x_1^2 + x_2^2 = 1$, with semi-axes $\frac{1}{2}$, 1 is mapped onto the unit circle.

EXAMPLE 2.18 (continued)

$$\mathbf{M'M} = \begin{bmatrix} 1 & 1 \\ 1 & 2 \end{bmatrix}, \text{ with eigenvalues } \tfrac{1}{2}(3 \pm \sqrt{5}).$$

The ellipse $x_1^2 + 2x_1x_2 + 2x_2^2 = 1$ is mapped onto the unit circle.

EXAMPLE 2.19 (continued)

$$\mathbf{M'M} = \begin{bmatrix} 0 & 0 \\ 0 & 2 \end{bmatrix}.$$

The locus $2x_2^2 = 1$, which is the line pair $x_2 = \pm 1/\sqrt{2}$, is mapped onto the pair of points $(1/\sqrt{2}, 1/\sqrt{2})$, $(-1/\sqrt{2}, -1/\sqrt{2})$.

EXAMPLE 2.20 (continued)
The line pair $3x_1 + 4x_2 = \pm 1$ in R^2 is mapped onto the pair of points ± 1 in R^1.

EXAMPLE 2.21 (continued)

$$\mathbf{M'M} = \begin{bmatrix} 1 & 0 \\ 0 & 2 \end{bmatrix}.$$

The ellipse $x_1^2 + 2x_2^2 = 1$ in R^2 is mapped onto a circle in R^3, the circle being the intersection of the sphere

$$y_1^2 + y_2^2 + y_3^2 = 1$$

and the plane

$$y_2 = y_3.$$

EXAMPLE 2.22 (continued)
The pair of planes $3x_1 + 4x_2 = \pm\sqrt{(1/14)}$ is mapped onto the pair of points $(1/\sqrt{14}, 2/\sqrt{14}, 3/\sqrt{14})$ and $(-1/\sqrt{14}, -2/\sqrt{14}, -3/\sqrt{14})$, which lie in the unit sphere in R^3.

EXAMPLE 2.23 (continued)
The spheroid $4x_1^2 + x_2^2 + x_3^2 = 1$ is mapped onto the unit sphere.

EXAMPLE 2.24 (continued)

$$\mathbf{x'M'Mx} = x_1^2 + x_2^2 = 1 \text{ is a circular cylinder.}$$

It maps onto the circle $y_1^2 + y_2^2 = 1$, $y_3 = 0$. We see a map *into* the unit sphere.

EXAMPLE 2.25 (continued)
The calculation is performed as in Example 2.24, but we have a map into R^2 instead of R^3 and the map may now be described as *onto*.

EXAMPLE 2.26 (continued)
An elliptic cylinder is mapped onto the unit circle.

EXAMPLE 2.27 (continued)
An elliptic cylinder is mapped onto the unit circle.

EXAMPLE 2.28 (continued)
The pair of planes

$$2x_1 + 2x_2 + x_3 = \pm 1/\sqrt{5}$$

in R^3 is mapped onto the pair of points

$$(1/\sqrt{5},\ 2/\sqrt{5}),\ (-1/\sqrt{5},\ -2/\sqrt{5})$$

in R^2; this is a map *into* the unit circle.

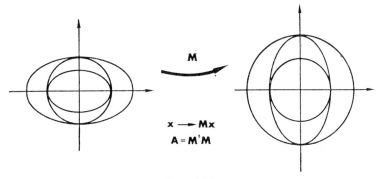

FIG. 2.13

Figure 2.13, an extension of Figure 2.12, may help to show what is involved. It shows maps of the unit sphere in the **x** space together with maps of the ellipsoids $\mathbf{x}'\mathbf{M}'\mathbf{M}\mathbf{x} = \lambda_{\max}$ and $\mathbf{x}'\mathbf{M}'\mathbf{M}\mathbf{x} = \lambda_{\min}$ which touch it externally and internally. It illustrates a typical case with non-zero eigenvalues; but this diagram should be considered in the light of the above examples.

The rank of a matrix was mentioned in Section 2.3. The idea of rank is very important in the development of matrix theory, and it is implicit in much of the work in the book. As one of our aims is to show as much as we can of the applications of matrices and linear algebra without developing theory for its own sake, and as rank is a notion which may not be familiar to some readers, we are dispensing with the notion as much as we can.

However, Examples 2.2 to 2.10 of Section 2.2 and Examples 2.15 to 2.28 illustrate the idea. In these examples consider the *image* under the map $\mathbf{x} \to \mathbf{M}\mathbf{x}$; that is to say consider the set of all vectors which arise as images under the transformation. In set notation this is the set $\{\mathbf{y}: \mathbf{y} = \mathbf{Mx}\}$. It is easy to show that this set is a vector space. (How?) As a vector space it has a certain dimension.

The rank of $\mathbf{M}$ is defined to be the dimension of this image space.

As a theoretical point, there is clearly the possibility of considering also the effect which the matrix $\mathbf{M}$ has on row vectors by the transformation $\mathbf{x} \rightarrow \mathbf{xM}$. ($\mathbf{x}$ now denotes a row vector.) As before, the image space, that is, the set of vectors $\{\mathbf{z}: \mathbf{z} = \mathbf{xM}\}$, has a certain dimension; and this dimension may be called the row rank, distinguishing it from the column rank which was defined earlier.

It is by no means immediately obvious that the row rank and the column rank of a matrix will always be the same, but it is possible to prove that this is so. Thus one may speak merely of the rank of a matrix, this being the same whether the underlying space is the space of rows or the space of columns.

An alternative way of formulating the idea of rank is to say that the statement 'the row rank of $\mathbf{M}$ is r' means that $\mathbf{M}$ possesses r linearly independent rows and no more. Column rank may be defined in the same way, and once again, for a detailed theoretical treatment, it is necessary to establish the result that row and column rank are the same.

References

1. ASSOCIATION OF TEACHERS OF MATHEMATICS, *Mathematical Reflections*, pp. 125–30, CUP (1970).
2. MIDLANDS MATHEMATICS EXPERIMENT, Harrap (1967).
3. MATTHEWS, G., *Matrices, I and II*, Arnold (1964).
4. MIRSKY, L., *An Introduction to Linear Algebra*, OUP (1955).
5. NEILL, H. and MOAKES, A. J., *Vectors, Matrices and Linear Equations*, Oliver and Boyd (1967).
6. SCHOOL MATHEMATICS PROJECT, *Book 3*, CUP (1967). (Also other books in the series.)
7. SCOTTISH MATHEMATICS GROUP, *Book 8, Modern Mathematics for Schools*, Blackie/Chambers (1969).

3
Mechanical Vibrations

This chapter is concerned with the theory of the vibrations of certain mechanical systems. The first problems to be considered concern systems of particles with a small number of degrees of freedom. By *number of degrees of freedom* is meant the number of independent parameters which are needed in order to specify the position of the system completely. After this we consider problems which are similar in a number of respects, but where the vibrating system is continuous and where in consequence the number of degrees of freedom is infinite.

Over the last few decades it has been normal practice in mathematics courses to solve these problems, which are very similar from a physical point of view, by quite different mathematical techniques, the discrete problems being solved by matrix algebra and the continuous problems by differential equations. In studying these theories it is easy to lose sight of the original physical problems, and if this happens then some of the similarities of terminology, such as *eigenvalue* and *eigenvector*, become difficult to reconcile in what may seem to the student two rather distinct branches of mathematics. This comes about because although early writers such as Lagrange, who was the great pioneer in this field, sought to treat the continuous problems as limiting cases of discrete ones, there are difficulties in doing so, and as the need was felt for more rigorous methods of proof the techniques of pure mathematics which were developed tended to pull apart what had originally been a unified field of enquiry.

Today the engineer who requires *explicit* solutions to these two kinds of problem—the modes of vibration of discrete and continuous mechanical systems—still needs to call on the techniques of matrix algebra on the one hand and differential equations on the other; although when computing he may cast his differential equations into finite difference form and consider the discrete case, as Lagrange did in the beginning. However, the development of the theory of linear spaces has bridged the gap and restored the original unity, although these ideas are still outside most elementary courses.

We seek to study the original vibration problems by means of the separate techniques, but to stress the similarities and so bring out the essential ideas of the theory of linear spaces, which enables the two groups of problems to be seen in a common conceptual framework. The presentation is intended for readers who already have some experience of classical mechanics, approached by traditional methods; but later on an alternative way is suggested, which might provide a short cut to the ideas for students whose knowledge of mechanics is only slight. Such students should not be discouraged by the amount of algebraic manipulation in the initial examples, because most of this manipulation is concerned with hacking away irrelevancies and reducing the problems to first-order approximations which may be handled with linear algebra. Once this has been done the procedures for solution are always essentially the same.

3.1 Systems of particles

EXAMPLE 3.1

A particle of unit mass is at the mid-point of a light elastic string, of unstretched length $2l_0$, whose ends are fastened to two points in a horizontal plane at a distance $2l$ apart. The modulus of the string is μ. All motion is in the horizontal plane (Fig. 3.1).

Find how the particle behaves if

(i) it is pulled aside by a force perpendicular to the string, of constant magnitude f;
(ii) it is forced by a periodic force $f = p \sin kt$;
(iii) it oscillates freely.

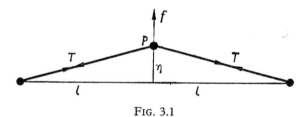

FIG. 3.1

If the displacement of the particle P is η then the extension of each half of the string is $\sqrt{(l^2 + \eta^2)} - l_0$.

By the definition of the modulus μ, the tension in the string is therefore

$$T = \mu\{\sqrt{(l^2 + \eta^2)} - l_0\}/l_0.$$

Resolving in the direction of the force f, the equation of motion is

$$f - 2T\eta/\sqrt{(l^2 + \eta^2)} = \ddot{\eta}.$$

This reduces to

$$f - 2\mu\eta\{l - l_0(1 + \frac{\eta^2}{l^2})^{-1/2}\}/ll_0 = \ddot{\eta}.$$

We now make a first-order approximation, valid when η is small. Indeed, *all our later work is going to be to first-order approximations.* Using the binomial theorem, we see that $\{1 + (\eta^2/l^2)\}^{-1/2} = 1 +$ terms in η^2 and terms of a higher order. Hence, ignoring terms in η^2 and of a higher order, the equation of motion becomes

$$f - a\eta = \ddot{\eta},$$

where $a = 2\mu(l - l_0)/ll_0$, and is constant.

If we have a static displacement $\ddot{\eta} = 0$ and

$$a\eta = f. \tag{3.1}$$

If $f = p \sin kt$ and we assume a response $\eta = y \sin kt$, (the assumption being justified only by the fact that we find a solution which fits!) then $\ddot{\eta} = -k^2 y \sin kt$, and

$$p \sin kt - ay \sin kt = -k^2 y \sin kt.$$

Hence

$$(a - k^2)y = p. \tag{3.2}$$

This gives a response η in phase with the perturbing force f, of an amplitude y which is proportional to the amplitude of the perturbing force, and which depends in a rather more complicated manner on the frequency of the perturbing force.

When $k^2 = a$ we cannot solve Eq. (3.2) for y, and we have to enquire more deeply if we want a detailed solution. This is a resonance condition.

Free oscillations arise when $p = 0$ and $k^2 = a$, for then we have an oscillation without any input force. Introducing a new piece of standard notation, the angular velocity of these free oscillations is given by the positive value of ω for which

$$\omega^2 - a = 0. \tag{3.3}$$

EXAMPLE 3.2

A particle of unit mass hangs at the end of a light string of length l. How does it behave if

(i) it is pulled aside by a constant horizontal force f;
(ii) it is forced by a periodic horizontal force $f = p \sin kt$;
(iii) if it swings freely?

The theory of the simple pendulum is well known.

Working as before, with an approximation valid for small displacements η,
$$a\eta = f, \qquad (3.1)$$
where this time $a = g/l$, g being the acceleration due to gravity. Once again for forced oscillations
$$(a - k^2)y = p. \qquad (3.2)$$
Also free oscillations occur with angular velocity given by the positive value of ω for which
$$\omega^2 - a = 0. \qquad (3.3)$$

EXAMPLE 3.3

Three particles, each of unit mass, are equally spaced at distances l on a light horizontal string fastened rigidly at the ends, resting on a smooth horizontal table (Fig. 3.2). How do the particles behave if

(i) they are displaced by constant horizontal forces, perpendicular to the string, of magnitudes f_1, f_2, f_3;
(ii) they are perturbed by periodic horizontal forces, perpendicular to the string, of magnitudes $p_1 \sin kt$, $p_2 \sin kt$, $p_3 \sin kt$;
(iii) they oscillate freely, moving perpendicular to the line of the string, in the horizontal plane?

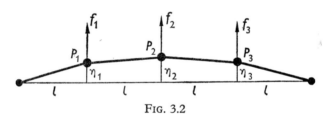

FIG. 3.2

Let the three particles P_1, P_2, P_3 have displacements η_1, η_2, and η_3.

For small displacements the lengths of the portions of the string are unaltered, neglecting terms in η_r^2 and higher orders; and so the tension T in the string is constant to this approximation also.

Resolving perpendicularly to the equilibrium position of the string for particle P_1, the two segments exert forces

$$-T\eta_1/l \quad \text{and} \quad T(\eta_2 - \eta_1)/l$$

respectively, measured in a direction of increasing η away from the equilibrium position. (Once again the small η approximations are used.)

If an external force f_1 is also acting on P_1, then the equation of motion is

$$f_1 - T\eta_1/l + T(\eta_2 - \eta_1)/l = \ddot{\eta}_1.$$

(Compare the corresponding steps in Example 3.1.)
Likewise for P_2 and P_3

and
$$f_2 - T(\eta_2 - \eta_1)/l - T(\eta_2 - \eta_3)/l = \ddot{\eta}_2,$$
$$f_3 + T(\eta_2 - \eta_3)/l - T\eta_3/l = \ddot{\eta}_3.$$

To reduce notation we will work with the physical constant T/l equal to unity, getting

$$\begin{aligned} f_1 - 2\eta_1 + \eta_2 &= \ddot{\eta}_1, \\ f_2 + \eta_1 - 2\eta_2 + \eta_3 &= \ddot{\eta}_2, \\ f_3 + \eta_2 - 2\eta_3 &= \ddot{\eta}_3. \end{aligned}$$

Adopting matrix notation, putting

$$\mathbf{A} = \begin{bmatrix} 2 & -1 & 0 \\ -1 & 2 & -1 \\ 0 & -1 & 2 \end{bmatrix}$$

these equations can be written

$$\mathbf{f} - \mathbf{A}\boldsymbol{\eta} = \ddot{\boldsymbol{\eta}}.$$

If the system is in equilibrium under the perturbing forces then $\ddot{\boldsymbol{\eta}} = 0$ and

$$\mathbf{A}\boldsymbol{\eta} = \mathbf{f}. \tag{3.1}$$

When $\mathbf{f} = \mathbf{p}\sin kt$ we determine whether it is possible to find a solution of the form $\boldsymbol{\eta} = \mathbf{y}\sin kt$. This means that $\ddot{\boldsymbol{\eta}} = -k^2\mathbf{y}\sin kt$ and the equations of motion can be satisfied by having

$$\mathbf{p}\sin kt - \mathbf{A}\mathbf{y}\sin kt = -k^2\mathbf{y}\sin kt,$$

or

$$(\mathbf{A} - k^2\mathbf{I})\mathbf{y} = \mathbf{p}. \tag{3.2}$$

Free oscillations can occur with angular frequency ω when

$$(\mathbf{A} - \omega^2\mathbf{I})\mathbf{y} = 0. \tag{3.3}$$

The calculation has been long, but the conclusions are encouraging because the equations (3.1), (3.2), (3.3) which answer the problems (i), (ii) and (iii) are of exactly the same form as in the first two examples, differing only in the fact that they are now matrix equations.

Over and over again we will see that the use of matrices enables methods of calculation in one variable to be extended to many variables.

Equation (3.3) is familiar ground; we need to find ω^2 and $\mathbf{y}$ which satisfy

$$\mathbf{A}\mathbf{y} = \omega^2\mathbf{y},$$

so it is a matter of finding eigenvalues and eigenvectors of $\mathbf{A}$. Proceeding as in Section 2.5 by expanding the characteristic determinant, or otherwise, we find that the eigenvalues of **A** are $2-\sqrt{2}$, 2, $2+\sqrt{2}$. These are the possible values of ω^2, since we have ω^2 instead of λ. A little calculation gives the corresponding eigenvectors.

$$\begin{aligned}\omega^2 &= 2-\sqrt{2}, & (1, \sqrt{2}, 1)'; \\ \omega^2 &= 2, & (1, 0, -1)'; \\ \omega^2 &= 2+\sqrt{2}, & (1, -\sqrt{2}, 1)'.\end{aligned}$$

It is very important to understand the physical significance of these quantities. There are three *normal modes* of vibration. A practical experiment will show that quite complex oscillations can take place in this comparatively simple physical system. The normal modes are especially simple oscillations in which the particles execute simple harmonic motions in phase with one another. The eigenvalues give the squares of the angular frequencies, and the eigenvectors give the relative displacements of the particles in the different normal modes. It will be remembered that the particles have displacements η_1, η_2, η_3 and that $\boldsymbol{\eta} = \mathbf{y} \sin \omega t$.

Figure 3.3 gives snapshots of the system vibrating in the three normal

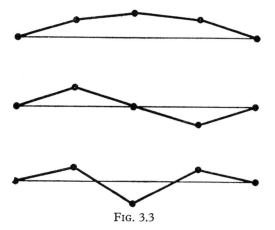

Fig. 3.3

modes. The reader who has any doubts about the practical realizability of the present theory would find it well worth while to thread some beads on a piece of elastic and conduct some experiments. The correspondence between algebra and physical experiment should be kept in mind throughout the whole of the discussion.

The three normal modes are particularly simple free vibrations. Linear sums of the normal modes are realizable physically. The normal modes are the simplest basis to take for the vector space of free vibrations.

Another example, an extension of Example 3.2, will lead to equations of the same form once again.

EXAMPLE 3.4

Three particles, each of mass m, are spaced at distances l on a light vertical string which hangs freely from a point of support (Fig. 3.4). All subsequent displacements are in a fixed vertical plane. How do the particles behave if

(i) they are displaced by constant horizontal forces of magnitudes f_1, f_2, f_3;
(ii) they are perturbed by periodic horizontal forces of magnitudes $p_1 \sin kt, p_2 \sin kt, p_3 \sin kt$;
(iii) they oscillate freely?

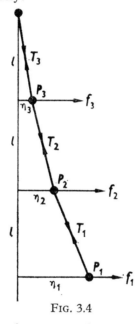

FIG. 3.4

If the tensions in the three parts of the string are T_1, T_2, T_3, then resolving vertically for the three particles, the net vertical forces on the three are respectively

$$T_1 \cos \theta_1 - mg,$$
$$-T_1 \cos \theta_1 + T_2 \cos \theta_2 - mg,$$
$$-T_2 \cos \theta_2 + T_3 \cos \theta_3 - mg;$$

if the segments are inclined to the vertical at angles $\theta_1, \theta_2, \theta_3$.

Resolving horizontally, if forces f_1, f_2, f_3 are imposed, then the net horizontal forces, to the right, are

$$f_1 - T_1 \sin \theta_1,$$
$$f_2 + T_1 \sin \theta_1 - T_2 \sin \theta_2,$$
$$f_3 + T_2 \sin \theta_2 - T_3 \sin \theta_3.$$

If the horizontal displacements of the particles are η_1, η_2, η_3, then
$$\eta_3 = l \sin \theta_3, \qquad \eta_2 - \eta_3 = l \sin \theta_2, \qquad \eta_1 - \eta_2 = l \sin \theta_1.$$

The equations are somewhat complicated, but we intend to work only to a first-order approximation to get a theory adequate enough to explain small vibrations of the system. Working to this order, tedious investigation of the detail, which we will not go into, shows that the vertical accelerations of the particles are of order θ^2, which we ignore, so the net vertical forces are zero and

$$T_1 = mg, \qquad T_2 = 2mg, \qquad T_3 = 3mg.$$

Equating the net horizontal forces to the masses multiplied by the accelerations

$$\begin{aligned} f_1 - T_1 \sin \theta_1 &= m\ddot{\eta}_1, \\ f_2 + T_1 \sin \theta_1 - T_2 \sin \theta_2 &= m\ddot{\eta}_2, \\ \text{and} \quad f_3 + T_2 \sin \theta_2 - T_3 \sin \theta_3 &= m\ddot{\eta}_3. \end{aligned}$$

Substituting the expressions for T_1, T_2, T_3 and $\sin \theta_1, \sin \theta_2$ and $\sin \theta_3$, working to the first order

$$\begin{aligned} f_3 - mg(\eta_1 - \eta_2) &= ml\ddot{\eta}_1, \\ f_2 + mg(\eta_1 - \eta_2) - 2mg(\eta_2 - \eta_3) &= ml\ddot{\eta}_2, \\ f_3 + 2mg(\eta_2 - \eta_3) - 3mg\eta_3 &= ml\ddot{\eta}_3. \end{aligned}$$

To reduce the number of symbols we may work with physical units such that $mg = ml = 1$, and these equations then become

$$\begin{aligned} f_1 - \eta_1 + \eta_2 &= \ddot{\eta}_1, \\ f_2 + \eta_1 - 3\eta_2 + 2\eta_3 &= \ddot{\eta}_2, \\ f_3 \phantom{{}+ \eta_1} + 2\eta_2 - 5\eta_3 &= \ddot{\eta}_3. \end{aligned}$$

Putting

$$\mathbf{A} = \begin{bmatrix} 1 & -1 & 0 \\ -1 & 3 & -2 \\ 0 & -2 & 5 \end{bmatrix},$$

we are back on familiar ground, as the equations of motion become

$$\mathbf{f} - \mathbf{A}\boldsymbol{\eta} = \ddot{\boldsymbol{\eta}},$$

as in the previous problem.

As before, if the system is in equilibrium under the perturbing forces then $\ddot{\boldsymbol{\eta}} = 0$ and

$$\mathbf{A}\boldsymbol{\eta} = \mathbf{f}. \tag{3.1}$$

If the perturbing forces are of the form $\mathbf{f} = \mathbf{p} \sin kt$ we find that there is a response $\boldsymbol{\eta} = \mathbf{y} \sin kt$, where

$$(\mathbf{A} - k^2\mathbf{I})\mathbf{y} = \mathbf{p}. \tag{3.2}$$

Free oscillations can occur with angular frequency ω when

$$(\mathbf{A} - \omega^2 \mathbf{I})\mathbf{y} = 0. \tag{3.3}$$

The numbers in the matrix $\mathbf{A}$ are quite small, and arise from a simple problem, but nevertheless the eigenvalues of $\mathbf{A}$ are not easy to calculate. The methods of Section 2.5 are hardly practical; it is necessary to adopt numerical methods based on Rayleigh's principle, which are described later in this chapter, or methods of the type described in Section 6.6. Computed by some such method it can be found that the eigenvalues and eigenvectors associated with Eq. 3.3 are

$$\omega_1^2 = 0\cdot416, \quad (0\cdot843, 0\cdot493, 0\cdot215)',$$
$$\omega_2^2 = 2\cdot29, \quad (-0\cdot528, 0\cdot683, 0\cdot505)',$$
$$\omega_3^2 = 6\cdot29, \quad (0\cdot102, -0\cdot540, 0\cdot836)'.$$

The corresponding modes of vibration are sketched in Fig. 3.5. These eigenvectors are normalized, i.e. the scale factor is chosen so that the sum of the squares of the components is unity.

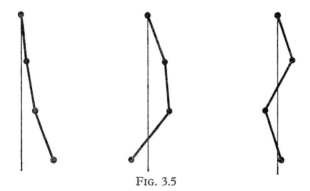

FIG. 3.5

It should now be clear that all of these apparently different physical problems are producing the same algebraic equations. From a purely mathematical point of view the problem is merely to solve Eq. (3.1) if we wish to know what displacements will be produced by an imposed set of constant forces; Eq. (3.2) if we wish to know the amplitudes of vibration which will be produced by an imposed set of periodic forces; and Eq. (3.3) if we wish to find the normal modes of vibration of the system.

Solving (3.1) involves inverting the matrix $\mathbf{A}$.
Solving (3.2) involves inverting the matrix $\mathbf{A} - k^2\mathbf{I}$.
Solving (3.3) involves finding the eigenvalues and eigenvectors of $\mathbf{A}$.
$\mathbf{A}$ is sometimes called the *stiffness matrix*.

The future developments are twofold. We hope to show further interconnections between these three problems; and we also intend to formulate some problems on differential equations in such terms that the correspondence with these matrix problems is made plain.

In the above problems we have encountered what might be termed the η-space, the f-space, the y-space and the p-space. These are *not* the same as the physical three-space in which the original problem was posed. If there had been four particles on each string then the η, f, y and p spaces would have been four-dimensional. The number of dimensions is equal to the number of particles. Furthermore, if each particle had two degrees of freedom this would double the number of dynamical unknowns, and so double the number of dimensions in the vector space which we use to describe the system. Three degrees of freedom to each particle would triple the number of dimensions.

A particular motion of the system corresponds to the motion of a single representative point in the η-space. This gives as it were a 'graph' of the motion. The normal modes of vibration are graphed by simple harmonic motions of the representative point along particular line segments through the origin in the y-space.

The eigenvectors corresponding to the normal modes may be normalized, and the work in Theorem 2.4 of Section 2.8 may be applied. This leads to a new co-ordinate system which is particularly well adapted to describing oscillations in a system such as this. These co-ordinates are called *normal co-ordinates*. The idea is simply that instead of describing displacements of the system in terms of η_1, η_2 and η_3 displacements are described in terms of the normal modes—a displacement being regarded as composed of so much of mode number one plus so much of mode number two, etc.

In some physical problems we may wish, say, to set the particles in motion with particular initial displacements *and* with particular initial velocities. This is possible because of the nature of the laws of dynamics. To deal with questions of this kind it is sometimes useful to double up the number of dimensions of our space again, to consider displacements of the form $\boldsymbol{\eta} = \mathbf{y} \cos kt$ also, and have axes corresponding to the velocities of the particles as well as corresponding to the displacements. Alternatively, where oscillations with different phases are concerned it is sometimes useful to use complex variables for co-ordinates in order to be able to handle both the amplitude and the phase of the oscillation with a single number.

These are complications which we will not pursue, our treatment being intended only as an introduction.

3.2 Solution of the general problems

The static problem is to determine $\boldsymbol{\eta}$ knowing $\mathbf{f}$, when

$$\mathbf{A}\boldsymbol{\eta} = \mathbf{f}. \tag{3.1}$$

This is simply a matter of introducing the inverse matrix, which for reasons which will emerge later we will call **G**. We have

$$\eta = \mathbf{G}f, \quad \text{where } \mathbf{G} = \mathbf{A}^{-1}.$$

G is sometimes called the *flexibility matrix*.

The matrices **A** and **G** map the η-space on to the f-space, and vice versa. **G** transforms a force vector into the corresponding displacement vector, and **A** transforms a displacement vector into a force vector. It is important to recognize just in what sense these things are vectors. If we had a problem with n particles on the string then these vectors would have n components; they would be n-dimensional vectors. We call them 'vectors' because two force vectors may be added according to the usual law of composition of vectors and not because we have ill-formulated ideas about representing forces by arrows.

The matrix **G** merits closer attention. What is the significance of the elements g_{rs}? If all the components of **f** are zero, except f_s, then $\eta_r = g_{rs}f_s$. In other words g_{rs} gives the displacement of particle r due to a unit force applied at point number s. (It is most important to remember this.)

This means that **G** can be found by experiment. Unit loads may be applied at the various points and the corresponding displacements can be observed. When these are tabulated we get the matrix **G**.

This makes possible a quite different approach to the whole problem. The examples in this chapter have been approached from the point of view of a reader with a fair background of mechanics as traditionally taught in the sixth forms of schools or in the first year of university. The calculations have involved many tedious details of a kind which was common in such courses but which are quite irrelevant to the main themes of this book. These details concern the processes of making a linear approximation to the problem. Many problems in industrial mathematics involve similar considerations; if it is reasonable to make a linear approximation then the hard work is largely over, or at least all further hard work can be transferred to computers because the computational routines appropriate to linear problems are well understood.

The student who lacks background in theoretical mechanics, as traditionally taught, can still appreciate this present range of vibrational problems if he is prepared to start by considering systems which are linear, that is in which the displacements are proportional to the forces imposed, and are additive in the sense that the displacements produced when two sets of forces act together are the sums of the displacements produced when the two sets act separately. (Once again the two essential ideas of linear spaces.) Assuming the linear response of the system to the imposed forces, and assuming Newton's second law of motion, that force is equal to mass multiplied by acceleration, then the whole of the present theory may be approached by way of matrix **G** rather than by the route adopted above. This might be a possible teaching strategy in the future.

Exercise 3.1
Calculate the matrices **G** for Examples 3.3 and 3.4.

In the examples we worked, **A** turned out to be a symmetric matrix. It is easy to show that the inverse of a symmetric matrix is symmetric and so **G** was necessarily symmetric as well. This means that $g_{rs} = g_{sr}$, which means in turn that a unit force applied at point s produces a displacement at point r which is equal to the displacement at point s produced by a unit force at point r. It is not obvious that this is always going to be so, but this property can be seen as a consequence of the principle of conservation of energy.

THEOREM 3.1
If energy is conserved then the flexibility matrix is symmetric.

Proof
For notational simplicity we will prove that $g_{12} = g_{21}$. The proof is similar for any other pair of terms. The method is to start with the undisplaced system, to impose a unit force at point 1 and then a unit force at point 2 and to calculate the energy change. We then start again, but impose the forces in reverse order. Equating the energies in the two cases produces the required result.

If a constant force f acts through a distance x then the work done is fx. If a force which varies linearly with x acts through a distance x then the work done is equal to the mean force multiplied by x. (This is the familiar rule of 'half stretching force times stretch'.)

A unit force acting at point 1 produces displacements g_{11} and g_{21} at points 1 and 2 respectively. The work done, which is stored as elastic energy in the system, is $\frac{1}{2}g_{11}$.

If now a unit force is imposed at point 2 it produces additional displacements g_{12} and g_{22} at points 1 and 2 respectively, and it does work $\frac{1}{2}g_{22}$. At the same time the first force acts through a further distance g_{12}, doing work g_{12}.

Hence imposing unit forces first at point 1 and then at point 2 the total energy stored is

$$\tfrac{1}{2}g_{11} + \tfrac{1}{2}g_{22} + g_{12}.$$

If the forces are imposed in reverse order the energy stored is

$$\tfrac{1}{2}g_{11} + \tfrac{1}{2}g_{22} + g_{21}.$$

Therefore, assuming that energy is conserved, $g_{12} = g_{21}$.

When periodic forces are applied and it is desired to calculate the amplitudes of the oscillations which result from applied forces of given amplitudes, it is necessary to solve Eq. (3.2).

$$(\mathbf{A} - k^2\mathbf{I})\mathbf{y} = \mathbf{p}. \qquad (3.2)$$

Now it is necessary to invert $\mathbf{A} - k^2\mathbf{I}$. In some circumstances this may be quite convenient, but if there is any question of being concerned with perturbing forces of a variety of frequencies then it may be better to say

$$(\mathbf{I} - k^2\mathbf{G})\mathbf{y} = \mathbf{G}\mathbf{p},$$

since $\mathbf{G} = \mathbf{A}^{-1}$. Then

$$\begin{aligned}\mathbf{y} &= \mathbf{G}\mathbf{p} + k^2\mathbf{G}\mathbf{y}, \\ &= \mathbf{G}\mathbf{p} + k^2\mathbf{G}(\mathbf{G}\mathbf{p} + k^2\mathbf{G}\mathbf{y}), \text{ etc.},\end{aligned}$$

leading to the series

$$\mathbf{y} = (\mathbf{I} + k^2\mathbf{G} + k^4\mathbf{G}^2 + k^6\mathbf{G}^3 + \ldots)\,\mathbf{G}\mathbf{p}.$$

This process may or may not converge, but it suggests the possibility of solving Eq. (3.2) by iteration. There is a further point of interest. We have seen that $\mathbf{G}$ can be found by *static* experiments, but the equation we have just derived gives the response of the system to periodic forces and this is a *dynamic* question. The answer to the static problem may be transferred to the dynamic one; and there is obviously practical importance in this, as it suggests, for example, how potentially dangerous resonances in vibrating structures can be calculated from static experiments.

The free oscillations (normal modes) are found from

$$(\mathbf{A} - k^2\mathbf{I})\mathbf{y} = 0, \qquad (3.3)$$

or

$$(\mathbf{I} - k^2\mathbf{G})\mathbf{y} = 0.$$

The angular frequencies of the normal modes are equal to the values of k for which $\mathbf{A} - k^2\mathbf{I}$, or $\mathbf{I} - k^2\mathbf{G}$, is singular.

Observe that complications arise if the matrix $\mathbf{A}$ is singular as $\mathbf{G}$ does not then exist. In this case $\mathbf{A}$ has a zero eigenvalue. This case is of physical importance, and flywheels mounted on a shaft which is free to rotate at the supports provide an example. We cannot go into details here, but the work by Prentis and Leckie [5] can be recommended. Other physical and mechanical applications of matrices can be found in Eisenschitz [2] and Heading [4].

Throughout this discussion we consider only mechanical systems, as these are the most familiar in terms of ordinary experience; but electrical systems may be treated in a precisely similar way, and the complete correspondence of the two theories is most instructive. Some of the ideas even apply to oscillations in biological populations or economic systems.

3.3 Oscillations in the continuous case

EXAMPLE 3.5

An elastic string of unit length is fastened to rigid supports at its extremities. Consider its behaviour when

(i) it is displaced by a distribution of forces $f(x)\, \delta x$;
(ii) it is acted on by a periodic force with a distribution $p(x)\, \delta x \sin kt$;
(iii) it oscillates freely.

This classic example was discussed by Lagrange. It sometimes seems to occupy an excessive space in mathematics courses, but it introduces principles which have far wider applications and it is the prototype of an extensive range of problems. Few practical problems have stimulated more, and deeper, pure mathematics than the problem of the vibrating string. As far back as the Greeks interest was shown in the numerical relations involved.

The first task is to establish the partial differential equation satisfied by the vibrating string when making transverse vibrations of small magnitude. It does not seem possible to derive this equation in a way which is both simple and rigorous, so we will give a few considerations which suggest a certain equation as a plausible starting point, then we will assume that this equation is an acceptable formulation of the physical problem, and then investigate solutions.

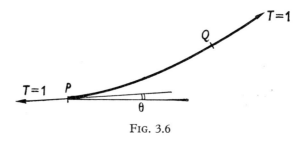

FIG. 3.6

We consider a string of length unity. The co-ordinate denoting the position on the string will be x, and the lateral displacement $\eta(x)$. We have $0 \leqslant x \leqslant 1$, and $\eta(0) = \eta(1) = 0$, since the ends of the string are fixed. Figure 3.6 shows an element of the string PQ. In the ensuing discussion only the transverse motion of the element is considered, any rotational and longitudinal motions being ignored as they are small quantities of the second order. (A fully detailed demonstration of this would be tedious.) This means that to the present order of approximation the tension along the string is constant. To reduce symbolism we will take it as unity.

The component of the tension in the string at P normal to the axis is $\sin \theta$. We are working to the first order in θ, with the approximations $\cos \theta = 1$, $\sin \theta = \tan \theta = \theta$, so to this accuracy the component is $\sin \theta = \tan \theta = d\eta/dx$.

The component at Q is $d\eta/dx$ evaluated at $x + \delta x$, which is

$$\frac{d\eta}{dx} + \frac{d}{dx}\left(\frac{d\eta}{dx}\right) \delta x$$

to the first order. (Later there will be differentiations with respect to t also; and if partial differential symbols are preferred for the differentiations with respect to x at this stage they may be used.)

Hence the resultant force on the element, acting away from the axis, is the difference of the two forces, namely

$$\frac{d}{dx}\left(\frac{d\eta}{dx}\right) \delta x = \frac{d^2\eta}{dx^2} \delta x.$$

We will imagine also that there is an externally applied force on the string $f(x)$. Working in this general way allows for a distributed force along the length of the string. This is not very likely in the case of, say, the vibrating strings on stringed instruments, but in the case of something like the cables on a suspension bridge this term can be very important. The external force on the element is then $f(x) \delta x$.

Once more avoiding unnecessary algebraic parameters we will assume that the string has unit mass per unit length, so the mass of the element is δx and its acceleration is $\ddot{\eta}$, a dot denoting differentiation with respect to t.

Equating the total force to the mass multiplied by the acceleration, the equation of motion for the element is

$$\frac{d^2\eta}{dx^2} \delta x + f(x) \delta x = \ddot{\eta} \delta x.$$

Using primes to denote differentiation with respect to x, the equation becomes

$$\ddot{\eta} - \eta'' = f(x), \qquad \eta(0) = \eta(1) = 0.$$

This is the standard equation for the motion of the string.

If the string is static, merely displaced by the applied force, the acceleration is zero everywhere and

$$-\eta'' = f(x). \qquad \eta(0) = \eta(1) = 0.$$

We can bring out the resemblance to the discrete problems by restricting attention to the set of functions which are

(i) defined over the interval $(0, 1)$;
(ii) vanish at $x = 0$ and $x = 1$;
(iii) possess second derivatives over the interval (there are actually technical reasons why it is better, in a rigorous treatment, to say *continuous* second derivatives—but this technicality will not concern us here).

These functions form a function space. This means merely that the set of them is closed with respect to addition and multiplication by a real scalar multiplier. This space is a space of infinite dimension.

Using $\mathscr{D}$ to denote the operator $-d^2/dx^2$, the equation to be solved is

$$\mathscr{D}\eta = f. \tag{3.1}$$

$\mathscr{D}$ is a differential operator which maps the η-space on to the f-space just as the matrix operator $\mathbf{A}$ did before. The formal similarity to the previous equation (3.1) is obvious, but what is the inverse operator $\mathscr{D}^{-1}$? We return to this question later.

The forced oscillations can be handled as follows. As before we consider the effect of imposed sinusoidal forces. We consider these because they turn out to be the simplest to handle, but they are an approximation to an ordinary shake, and there are more advanced methods available by which the responses to forces which vary with time in a more complicated manner can be deduced from the responses to forces which vary with time sinusoidally. If the disturbing force is

$$f(x, t) = p(x) \sin kt$$

then, assuming that we may find a solution

$$\eta(x, t) = y(x) \sin kt,$$

substituting into the original equation of motion of the string

$$-k^2 y - y'' = p, \quad \text{or} \quad (\mathscr{D} - k^2)y = p. \tag{3.2}$$

y and p are functions of x, and $\mathscr{D} - k^2$ is the operator $-d^2/dx^2 - k^2$. The solution of this equation will be discussed later.

The free vibrations arise as solutions of

$$(\mathscr{D} - k^2)y = 0.$$

Once again these are vibrations which are self maintaining without any externally applied force. They arise only for particular values of k—the resonant frequencies.

Bringing this into line with our previous notation, the normal modes occur as solutions of the equation

$$(\mathscr{D} - \omega^2)y = 0. \tag{3.3}$$

To agree completely with previous notation y should be printed in bold-face type, but it seems more in conformity with usage to continue to use italic.

Equation (3.3) has eigenfunctions and eigenvalues as we will shortly see, and these give the normal modes of vibration of the string. In more normal calculus notation we have to solve

$$y'' + \omega^2 y = 0, \quad y(0) = y(1) = 0.$$

The first equation is the familiar differential equation for simple harmonic motion, and it has solution

$$y = a \sin \omega x + b \cos \omega x,$$

where a and b are arbitrary constants. If we are to fit the boundary condition $y = 0$ when $x = 0$, we must have $b = 0$. Since $y(1) = a \sin \omega$ we can only fit the boundary condition $y = 0$ when $x = 1$ by arranging that

$$\sin \omega = 1.$$

This means that

$$\omega = n\pi, \quad n = 1, 2, 3, \ldots \, .$$

It is at this point that the eigenvalues arise. The angular frequencies of the normal modes of the string are $n\pi$, for integer n, and the corresponding eigenfunctions are $y = \sin n\pi x$. These give the familiar snapshots of the string vibrating in various modes, producing overtones or harmonics (Fig. 3.7).

FIG. 3.7

To stress the correspondence with the matrix case we may rewrite Eq. (3.3) as

$$\mathscr{D} y = \lambda y, \tag{3.3}$$

and repeat that $\mathscr{D}$ denotes the operator $-d^2/dx^2$, which operates on a particular space of functions—functions which are defined over the interval $0 \leqslant x \leqslant 1$ and possess continuous second derivatives over the interval. We have seen that functions y giving a solution can only be found for certain values of λ, called the eigenvalues. In this case $\lambda = \omega^2$ and the eigenvalues of λ are $n^2\pi^2$, $n = 1, 2, 3, \ldots$. Zero is not an eigenvalue, because this leads to $y(x) = 0$, which is a solution of the equation, but not a solution of any interest. To each eigenvalue there corresponds an eigenfunction. As with matrices, eigenfunctions are determined apart from a scalar multiplier. Any multiple of $\sin n\pi x$ satisfies the differential equation.

Equations (3.1), (3.2) and (3.3) in this example should be compared with the corresponding equations in the earlier examples.

EXAMPLE 3.6

This example is the continuous analogue of Example 3.4. It was discussed by D. Bernoulli round about 1732.

A uniform chain of unit length hangs vertically from a fixed point (Fig. 3.8). It is displaced in a vertical plane. Consider its behaviour when

(i) it is displaced by a distribution of forces $f(x)\,\delta x$;
(ii) it is acted on by a periodic force with distribution $p(x)\,\delta x \sin kt$;
(iii) it swings freely.

The only fresh difficulty is to form the equation of motion of an element of the chain.

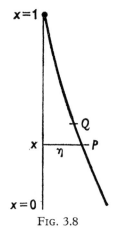

FIG. 3.8

Let the co-ordinate x denote the position on the chain measured from the free end. Let the lateral displacement of the chain at point x be $\eta(x)$. To reduce symbolism we will take the mass per unit length of the chain as unity.

Working as in the previous example, resolving vertically for the element shows that the tension in the chain is, neglecting terms of second and higher orders (a very tedious matter to do in detail),

$$T(x) = gx.$$

This says simply that the tension at each point has to support the weight of the chain below the point.

Resolving perpendicular to the undisplaced position of the chain, the net force on the element due to the tensions at P and Q is $(d/dx)(gx\eta')\delta x$. (Compare the corresponding step in the previous example.) To this must be added the exterior force $f(x)\,\delta x$.

Equating the total force on the element of chain to its mass multiplied by its acceleration we get

$$\frac{d}{dx}(gx\eta')\,\delta x + f(x)\,\delta x = \ddot{\eta}\,\delta x,$$

which means that

$$\ddot{\eta} = \frac{d}{dx}(x\eta') + f(x),$$

if we choose units to make the gravitational constant unity. Again, we do not propose to discuss the formulation of this equation in detail; we will assume that it is an acceptable formulation of the physical problem and consider its mathematical solution.

If the chain is static we have $\ddot{\eta} = 0$ and

$$-\frac{d}{dx}\left(x\frac{d\eta}{dx}\right) = f.$$

This time putting

$$\mathscr{D} = -\frac{d}{dx}\left(x\frac{d}{dx}\right)$$

this equation may be written

$$\mathscr{D}\eta = f, \tag{3.1}$$

and the problem is now expressed in the same terms as before. $\eta(x)$ is to be found in the space of functions which are

(i) defined over $(0, 1)$,
(ii) vanish at $x = 1$;
(iii) are twice (continuously) differentiable over the interval.

It seems at first sight that there is no boundary condition at $x = 0$, but it turns out that the differential equation has solutions which tend to infinity as x tends to zero, and these will obviously not suit this problem. So there is in fact the further condition

(iv) $\eta(x)$ is bounded at $x = 0$.

The differential operator $\mathscr{D}$ maps the space of displacements onto the space of perturbing forces. Of course, these are spaces of infinite dimension.

Just as before, if there is a perturbing force $f(x) = p(x) \sin kt$, and it produces a response $\eta(x) = y(x) \sin kt$,

$$(\mathscr{D} - k^2)y = p; \tag{3.2}$$

and the normal modes are given by

$$(\mathscr{D} - \omega^2)y = 0. \tag{3.3}$$

These continuous problems are now expressed in the same form as the discrete problems, although the methods of getting explicit solutions are quite different in the two cases, at least for the moment.

As an example, consider Eq. (3.1) when the chain is displaced by a uniform force—say as a result of a steady wind blowing past it. Equation (3.1) becomes

$$-\frac{d}{dx}\left(x\frac{d\eta}{dx}\right) = c.$$

Hence

$$-x\frac{d\eta}{dx} = cx + c_1,$$

$$\frac{d\eta}{dx} = -c - \frac{c_1}{x},$$

and

$$\eta = -cx - c_1 \log x + c_2.$$

If we seek solutions which are finite at $x = 0$ then $c_1 = 0$. When $x = 1$, $\eta = 0$, so $c_2 = c$. Hence the displacement is given by

$$\eta = c(1 - x);$$

that is to say the chain is straight and slanting. Equation (3.3) may be written as

$$xy'' + y' + \omega^2 y = 0, \quad y(0) \text{ finite}, \quad y(1) = 0.$$

It is not possible to find a solution of this equation in terms of elementary functions, but a solution can be obtained as a power series. Let us see if the coefficients can be chosen so that the power series

$$y = a_0 + a_1 x + a_2 x^2 + \ldots + a_r x^r + \ldots$$

satisfies the differential equation.

Assuming that it is valid to differentiate term by term

$$y' = a_1 + 2a_2 x + 3a_3 x^2 \ldots + r a_r x^{r-1} + \ldots$$

and

$$y'' = 2a_2 + 6a_3 x \ldots + r(r - 1)a_r x^{r-2} + \ldots \;.$$

It is now necessary to multiply these three expressions by ω^2, 1 and x respectively and add. This gives

$$xy'' + y' + \omega^2 y = (a_0\omega^2 + a_1) + (a_1\omega^2 + 4a_2)x + (a_2\omega^2 + 9a_3)x^2 + \\ + \ldots + [a_r\omega^2 + (r+1)^2 a_{r+1}]x_r + \ldots .$$

The equation is satisfied if all of the coefficients on the right-hand side vanish. This requires

$$a_1 = -a_0\omega^2,$$
$$a_2 = -\tfrac{1}{4}a_1\omega^2 = \tfrac{1}{4}a_0\omega^4,$$
$$a_3 = -\tfrac{1}{9}a_2\omega^2 = \tfrac{1}{4}\cdot\tfrac{1}{9}a_0\omega^6,$$
$$\cdot \quad \cdot \quad \cdot \quad \cdot \quad \cdot \quad \cdot$$

and

$$a_{r+1} = -a_r\omega^2/(r+1)^2.$$

This gives in general, if we put $a_0 = 1$,

$$a_r = (-1)^r \omega^{2r}/(r!)^2$$

and the series required is

$$y = 1 - \omega^2 x + \omega^4 x^2/(2!)^2 - \omega^6 x^3/(3!)^2 + \ldots .$$

It can be shown that this series converges for all values of x. Any multiple of it satisfies the differential equation.

A related series

$$1 - \left(\frac{x}{2}\right)^2 + \frac{1}{(2!)^2}\left(\frac{x}{2}\right)^4 - \frac{1}{(3!)^2}\left(\frac{x}{2}\right)^6 + \ldots$$

is called the Bessel function $J_0(x)$, and so the series we have derived is the Bessel function $J_0(2\omega\sqrt{x})$. Bessel functions are extensively tabulated, and may be regarded as 'known'.

The eigenvalues and eigenfunctions come in when the boundary conditions are considered. In fact the requirement that our solution should be bounded near $x = 0$ has already been accommodated as there is in fact a second, independent solution of the differential equation which cannot be found by the method of solution in series (in the simple form in which we have used it) and which tends to infinity as x tends to zero.

The other boundary condition is $y(1) = 0$, and this requires

$$J_0(2\omega) = 0.$$

The values of argument for which the Bessel function J_0 is zero may be found in books of tables, and the first few values are

$$2\omega = 2\cdot 405,\ 5\cdot 520,\ 8\cdot 654,\ 11\cdot 79,\ 14\cdot 93,\ \text{etc.}$$

This gives the eigenvalues of ω for this problem. There is an infinite number of eigenvalues and there is no general formula for the nth eigenvalue. If the nth eigenvalue is denoted by ω_n then the corresponding eigenfunction is $J_0(2\omega_n\sqrt{x})$. Once again, this may be computed from tables. These functions give snapshots of the oscillations in the various normal modes, and sketches of the first few cases are shown in Fig. 3.9.

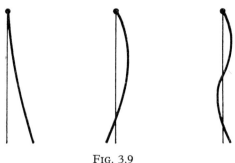

FIG. 3.9

The next example has interesting physical implications and it introduces a more general type of boundary condition.

EXAMPLE 3.7

Consider the vibrating string in Example 3.5, but with non-rigid pegs. In practice the pegs may be loose in a variety of ways, but as a comparatively simple case to investigate we will consider what happens when the peg at $x = 1$ yields elastically. That is, we will assume that during the motion there is no movement in the direction of the x-axis, but there is some movement in the η-direction produced by the component of the tension in the string, as it vibrates. We assume further that this movement of the supporting peg is controlled by an elastic restoring force—a force whose magnitude is proportional to the displacement. Of course both pegs might suffer from this type of looseness, but by considering one we will be able to indicate sufficiently clearly the fresh problems which arise. We neglect the mass of the peg; but it may be shown that taking this into consideration leads to a boundary condition of the same type, although the coefficients appearing are more complicated.

The situation at the peg at $x = 1$ is shown in Fig. 3.10.

If the tension in the string is T, then the component normal to the x-axis is $T \sin \phi$, and working as always with the first-order approximations for which $\sin \phi = \phi = \tan \phi$ this component is

$$-T\left(\frac{d\eta}{dx}\right)_{x=1}.$$

With our assumptions about the elastic restoring force on the peg, the force arising from the component of the tension in the string is balanced by the restoring force $-c\eta$, where c is constant. Hence the boundary condition is

$$c\eta = T\frac{d\eta}{dx}, \quad x = 1;$$

and this may be written in the form

$$\alpha\eta + \beta\eta' = 0, \quad x = 1.$$

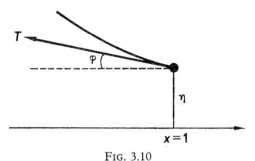

Fig. 3.10

Working exactly as in Example 3.5, putting $\eta(x, t) = y(x) \sin kt$ and denoting the values of k associated with the normal modes by ω, the equations for $y(x)$ are

$$y'' + \omega^2 y = 0 \quad \text{and} \quad y(0) = 0;$$

but this time we have

$$\alpha y + \beta y' = 0, \quad x = 1.$$

We may note in passing that many other physical problems, such as those arising in the theory of heat conduction and in theories of wave propagation of all kinds, also lead to this type of boundary condition.

As before, the differential equation and the boundary condition at $x = 0$ are satisfied by all functions of the form $y = \sin \omega x \times$ constant. The new boundary condition at $x = 1$ imposes the requirement

$$\alpha \sin \omega + \beta\omega \cos \omega = 0,$$

and so the eigenvalues are given as the roots of the equation

$$\tan \omega = -\frac{\beta}{\alpha}\omega.$$

This can be solved graphically by drawing Fig. 3.11. Whilst α/β may have either sign in the cases with which we are concerned it is positive for physical reasons, and the graph is drawn for this case. Inspection of the graph shows that all of the eigenvalues are reduced, the lowest value by a small amount and large values by approximately $\frac{1}{2}\pi$. Indeed for large eigenvalues instead of $\omega_n = n\pi$ we have

$$\omega_n = (n - \tfrac{1}{2})\pi \text{ approximately.}$$

This means that if the peg is loose the pitch is lowered, and also that the overtones are no longer exact multiples of the lowest frequency, which impairs the musical quality of the sound. The eigenfunction corresponding to the nth eigenvalue ω_n is

$$y = \sin \omega_n x,$$

and it cannot be expressed any more explicitly in terms of n than this.

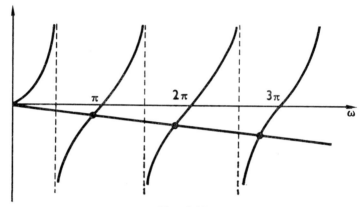

Fig. 3.11

So far there remains much to explain about the continuous cases. We have not solved Eq. (3.2) in any example, and whilst the use of the operator $\mathscr{D}$ notation has enabled the problems to be formulated in a way that is completely analogous to the discrete problems, the methods of solution which we have employed show no resemblance, and we have nothing to correspond to the matrix **G** which was the inverse of **A**. Is there any operator $\mathscr{G}$ (say) which can be regarded as inverting $\mathscr{D}$ as **G** inverted **A**?

The way on is indicated by remembering the physical significance of **G**. **G** was obtainable experimentally. It gives the displacement at point r when there is a unit load at point s. A similar experiment could be performed in Examples 3.5 and 3.6. We may place a unit load at point ξ and determine the displacement produced at point x. The mathematical solution comes by solving the equations which correspond to the physical experiment.

3.4 Green's functions

EXAMPLE 3.5 (continued)

The uniform string in Example 3.5 is pulled aside by a unit force at P, $x = \xi$. What is the resulting displacement?

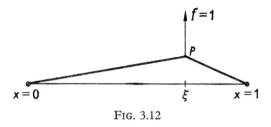

FIG. 3.12

The original differential equation now applies over each segment of the string, but *not* at the point $x = \xi$ where the function $\eta(x)$ is not differentiable. At this point there is static equilibrium, determined by the usual methods with three concurrent forces.

$\eta'' = 0$ has general solution $\eta = ax + b$.

The solutions which vanish at $x = 0$ are of the form

$$\eta = a_1 x, \quad 0 \leqslant x \leqslant \xi.$$

The solutions which vanish at $x = 1$ are of the form

$$\eta = a_2(1 - x), \quad \xi \leqslant x \leqslant 1.$$

If these solutions are equal when $x = \xi$, then $a_1 \xi = a_2(1 - \xi)$.

The condition for static equilibrium at $x = \xi$ can be obtained by resolving perpendicular to the undisplaced position of the string and equating the sum of the components of the tensions in the two parts of the string to the imposed force. We saw earlier that for the small displacements we are assuming that the tension remains unity throughout the string, and we also used the approximation $\sin \theta = \tan \theta = \eta'$ where θ is the inclination of the tangent to the string to the axis.

Hence the component of the tension in the right-hand portion of the string perpendicular to the axis, in the direction of η increasing is

$$\frac{d\eta}{dx} \text{ evaluated at the right of } P, \text{ which is } -a_2,$$

and the component in the left-hand portion of the string is

$$-\frac{d\eta}{dx} \text{ evaluated at the left of } P, \text{ which is } -a_1.$$

(The sign may require careful checking.) Writing this in the form which is useful in more general circumstances for later use, the total effect of the tensions in the two segments of string is

$$\operatorname*{Lt}_{h \to 0} [\eta'(\xi + h) - \eta'(\xi - h)].$$

This may be denoted more briefly as

$$\operatorname*{Lt}_{h \to 0} [\eta'(x)]_{\xi - h}^{\xi + h}$$

or more briefly still as

$$[\eta']_{\xi - 0}^{\xi + 0}.$$

This is equal to the unit force applied at P, and taking care of the sign we have

$$[\eta']_{\xi - 0}^{\xi + 0} = -1, \quad \text{or in this case} \quad a_1 + a_2 = 1.$$

Taken with the earlier equation we have $a_1 = 1 - \xi$ and $a_2 = \xi$. Hence the form of the string with this unit-perturbing force is

$$\begin{aligned} \eta(x) &= (1 - \xi)x, & 0 \leqslant x \leqslant \xi, \\ &= (1 - x)\xi, & \xi \leqslant x \leqslant 1. \end{aligned}$$

This function, which requires the two separate equations to specify it over the different ranges, we will denote by $G(x, \xi)$. x is the variable denoting a typical point on the string, ξ is the parameter which is the co-ordinate of the point at which the perturbing force is applied.

G is called the *Green's function* associated with the problem. We may refer to the matrix **G** in the earlier examples as the *Green's matrix* but this terminology is not very often used.

Having found G, the profile of the string for other more general displacing forces may then be calculated. It is a physically plausible argument to say that if a force $f(\xi) \, \delta\xi$ acts on each element of the string $(\xi, \xi + \delta\xi)$, then summing and proceeding to the limit we would expect the displacement of the string to be

$$\eta(x) = \int_0^1 G(x, \xi) f(\xi) \, d\xi.$$

Fuller justification for this will be given in Section 5.5.

In saying this we assume that the displacements produced by different forces at different places are *additive*. This is a direct consequence of the *linearity* of the equation, and it is linear because we have taken a linear approximation. It would ultimately be an experimental matter to decide whether or not our solution describes reality with sufficient accuracy to be of practical use.

Before commenting further, let us apply the same ideas to Example 3.6.

EXAMPLE 3.6 (continued)

The chain in Example 3.6 is drawn aside by a unit force at the point $x = \xi$ (Fig. 3.13). What is the resulting displacement?

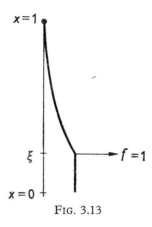

FIG. 3.13

The general solution of

$$-\frac{d}{dx}\left(x\frac{d\eta}{dx}\right) = 0$$

is given by

$$x\eta' = c,$$
$$\eta' = c/x,$$
$$\eta = c \log x + c_1.$$

For the two sections of the chain, fitting the constants to the boundary conditions we have

$$\eta = c_1, \qquad 0 \leqslant x \leqslant \xi,$$
$$\eta = c \log x, \qquad \xi \leqslant x \leqslant 1.$$

For the two solutions to match at $x = \xi$

$$c \log \xi = c_1.$$

For equilibrium at $x = \xi$ the appropriate condition this time is

$$[x\eta']_{\xi-0}^{\xi+0} = -1, \qquad \text{so } c = -1.$$

The displacement of the string is

$$\eta = G(x, \xi) = -\log \xi, \qquad 0 \leqslant x \leqslant \xi,$$
$$= -\log x, \qquad \xi \leqslant x \leqslant 1.$$

This is the Green's function for the present problem. Again, on physical grounds, if there is a distributed force with $f(\xi)\,\delta\xi$ acting on the element $(\xi, \xi + \delta\xi)$ we might expect that the displacement would be

$$\eta = \int_0^1 G(x, \xi) f(\xi)\, d\xi. \tag{3.4}$$

Again, fuller justification will be given for this formula in Section 5.5. What is now being said is that the solution of

$$\mathscr{D}\eta = f \tag{3.1}$$

in the *general* case is given by Eq. (3.4), where $G(x, \xi)$ is the solution of Eq. (3.1) in the *special* case when f is a point force applied at $x = \xi$. We may rewrite Eq. (3.4) as

$$\eta = \mathscr{G}f,$$

where $\mathscr{G}f$ denotes $\int_0^1 G(x, \xi) f(\xi)\, d\xi$. We can now regard $\mathscr{G}$ as the operator which is the inverse of the operator $\mathscr{D}$, as the matrix $\mathbf{G}$ was the inverse of the matrix $\mathbf{A}$. A detailed justification of all this, with full rigour, is the subject of specialized treatises. We will content ourselves for the moment with the plausible physical argument. Incidentally, this argument has been applied for very many years in school physics, most obviously in potential theory.

The potential at (x, y, z) due to a unit charge at $(\xi, \eta, \zeta,)$ is

$$V = \frac{1}{r} = \{(x - \xi)^2 + (y - \eta)^2 + (z - \zeta)^2\}^{-1/2}. \tag{3.5}$$

If there is a space distribution of charge of density $\rho(\xi, \eta, \zeta)$, then the potential at (x, y, z) is the volume integral

$$V(x, y, z) = \int_{\text{Vol}} \frac{\rho(\xi, \eta, \zeta)\, d\xi\, d\eta\, d\zeta}{\{(x - \xi)^2 + (y - \eta)^2 + (z - \zeta)^2\}^{1/2}}. \tag{3.6}$$

We may put this into more learned terms. The potential $V(x, y, z)$ satisfies Poisson's equation

$$\mathscr{D}V = \rho, \tag{3.7}$$

where

$$\mathscr{D} = -\frac{1}{4\pi}\left(\frac{\partial^2}{\partial x^2} + \frac{\partial^2}{\partial y^2} + \frac{\partial^2}{\partial z^2}\right).$$

Equation (3.6) gives the general solution of Eq. (3.7) utilizing (3.5); and (3.5), the familiar potential due to a unit charge, is a Green's function. It is the solution of Eq. (3.7) for the case when the distribution of charge $P(x, y, z)$ on the right-hand side is a unit charge at the single point (ξ, η, ζ).

This method is employed intuitively many times over in school physics. For example, when the potential due to a solenoid is obtained by integrating the potential due to a single turn over a range equal to the length of the coil.

Exercise 3.2
Find the Green's function for the problem in Example 3.7.

So far we have covered the solution of Eq. (3.1)—the static problem. What about Eqs. (3.2) and (3.3)?
Equation (3.2) was

$$(\mathscr{D} - k^2)y = p.$$

Two lines of attack are open: we may calculate a Green's function $\Gamma(x, \xi)$ for this (dynamic) system, and proceed as before. (Let us call this Method 1); or we may be guided by the matrix case and attempt an iterative solution, using the static Green's function $G(x, \xi)$ already found (Method 2).

Method 1
In the case of the string we have to solve

$$y'' + k^2 y = 0, \quad 0 < x < 1, \quad \text{with} \quad y(0) = y(1) = 0;$$

except at the point $x = \xi$, where there is an applied force $f = \sin kt$. This force is counterbalanced by the tensions in the string to the left and right at $x = \xi$, which, as in Example 3.5, amounts to the condition $[y']_{\xi-0}^{\xi+0} = -1$.

We will omit the details of the calculation, which we do not require subsequently, but the solution is interesting because of its various physical implications.

$$y = \Gamma(x, \xi) = \frac{\sin kx \sin k(1 - \xi)}{k \sin k}, \quad 0 \leqslant x \leqslant \xi,$$

$$= \frac{\sin k\xi \sin k(1 - x)}{k \sin k} \quad \xi \leqslant x \leqslant 1.$$

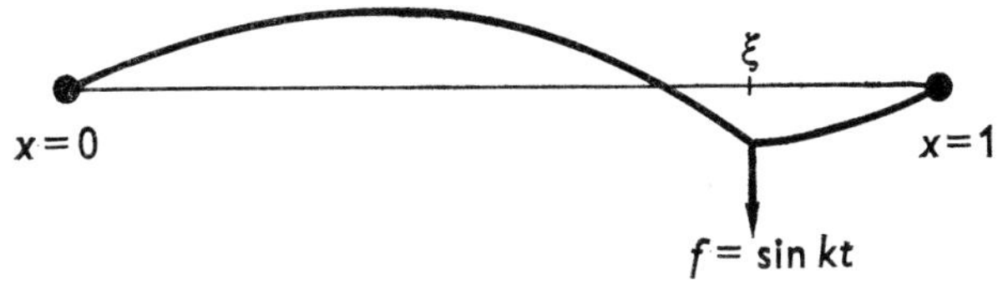

FIG. 3.14

These equations give a snapshot of the string when it is shaken at the point $x = \xi$ with a periodic force $\sin kt$ of unit amplitude (Fig. 3.14). The response depends on k, and there are some interesting special cases.

If $k = n\pi$ there is resonance and the expressions become infinite.

If $k \neq n\pi$ but $k\xi = m\pi$ then part of the string vibrates, the rest is stationary, and the point at which the force is applied does not move!

By the previous physical argument it is plausible that the solution of Eq. (3.2) in the general case of a distributed force whose amplitude at x is $p(x)$ should be

$$y = \int_0^1 \Gamma(x, \xi) \, p(\xi) \, \mathrm{d}\xi. \tag{3.8}$$

Method 2

Equation (3.2) was

$$(\mathscr{D} - k^2)y = p. \tag{3.2}$$

Rearranging,

$$\mathscr{D}y = p + k^2 y$$

This is a problem like Eq. (3.1), save that we have $p + k^2 y$ instead of p on the right-hand side. Proceeding formally

$$\begin{aligned} y &= \mathscr{G}(p + k^2 y), \\ &= \mathscr{G}p + k^2 \mathscr{G}y, \\ &= \mathscr{G}p + k^2 \mathscr{G}(\mathscr{G}p + k^2 y), \end{aligned}$$

and so on, until

$$y = \mathscr{G}p + k^2 \mathscr{G}^2 p + k^4 \mathscr{G}^3 p + \cdots$$

We might hope that the process on the right-hand side converges to the solution we seek. A full justification for this is beyond the scope of this book, but there are interesting implications, because if this process can be justified, like the similar process with matrices in Section 3.2 it enables the response of the system to periodic forces to be calculated from static experiments.

We have no general methods for finding the eigenvalues and eigenfunctions of our differential operators

$$\mathscr{D}y = \omega^2 y \tag{3.3}$$

(with the appropriate boundary conditions) comparable to the general methods available with matrices, and it would clearly be a great step forward if any such general methods could be found. The Green's function helps to some extent; $\mathscr{G}$ being in a certain sense the inverse operator of $\mathscr{D}$, Eq. (3.3) is equivalent to

$$y = \omega^2 \mathscr{G}y. \tag{3.3a}$$

But at our present state of knowledge this is hardly an improvement, because we have no way of computing the eigenvalues and eigenfunctions of any general form of Eq. (3.3a). It turns out that as (3.3a) is an equation involving an integral operator it is in some ways easier to work with, at least for purposes of numerical computation, than Eq. (3.3) which involves a differential operator; but this only becomes clear when further techniques are available. Some possible lines of attack are opened up by the next section.

3.5 Energy and Rayleigh's principle

In the mechanical examples discussed in this chapter no attention was given to the energy of the vibrating systems. This energy appears in two forms—potential and kinetic. The potential energy arises from the work done against the elastic restoring forces, and the kinetic energy is the energy which the system has as a result of the velocities of the particles. We will consider first the potential and kinetic energies involved in a simple harmonic motion. That is to say, we first consider a system with one degree of freedom, and it is worth considering it in detail because certain matters which are so simple as to seem hardly worth mentioning generalize to systems with n degrees of freedom, and there the properties are by no means so obvious.

Examples 3.1 and 3.2 both illustrate the first points which have to be made. In each of these examples Eq. (3.1) showed that for a small displacement η the restoring force was $a\eta$. The potential energy V stored in the system with this displacement is given by

$$V = \int_0^\eta a\eta \, d\eta = \tfrac{1}{2}a\eta^2.$$

Alternatively the formula 'half stretching force times stretch' may be applied, leading to the same answer.

If the particle is performing a simple harmonic motion $\eta = y \sin kt$ then the potential energy at any instant is

$$V = \tfrac{1}{2}ay^2 \sin^2 kt.$$

The kinetic energy of the particle at the same instant, remembering that it has unit mass, is

$$T = \tfrac{1}{2}\dot\eta^2 = \tfrac{1}{2}k^2y^2 \cos^2 kt.$$

The motion may be analysed in various ways. The total energy is given by the expression

$$\tfrac{1}{2}ay^2 \sin^2 kt + \tfrac{1}{2}k^2y^2 \cos^2 kt,$$

and this can only be constant if $a = k^2$, and this leads to the equation for the natural period of oscillation which was derived before. Another point of view may be taken.

$$\int_0^{2\pi/k} \sin^2 kt \, dy = \int_0^{2\pi/k} \cos^2 kt \, dt = \tfrac{1}{2},$$

and so the mean potential energy is

$$\bar{V} = \tfrac{1}{4}ay^2,$$

and the mean kinetic energy is

$$\bar{T} = \tfrac{1}{4}k^2 y^2.$$

It follows that the mean kinetic and potential energies are equal if and only if the particle is oscillating at its natural period. This property of simple harmonic motion, perhaps insufficiently stressed in elementary texts, is sufficient to characterize the natural period—that is to say if the property is accepted as a principle it derives the natural frequency. This is important because this principle can also be employed in vibration problems with many degrees of freedom.

In Examples 3.3 and 3.4 the displacements η and the displacing forces $\mathbf{f}$ are related by Eq. (3.3)

$$\mathbf{A}\eta = \mathbf{f}. \tag{3.3}$$

If the force f_1 acts through the distance η_1 then the potential energy stored is, as before $\tfrac{1}{2}\eta_1 f_1$. The total potential energy stored is

$$\tfrac{1}{2}(\eta_1 f_1 + \eta_2 f_2 + \eta_3 f_3) = \tfrac{1}{2}\eta'\mathbf{f}.$$
$$= \tfrac{1}{2}\eta'\mathbf{A}\eta,$$

since $\mathbf{f} = \mathbf{A}\eta$.

If $\eta = \mathbf{y} \sin kt$, working as before, the mean potential energy works out to be

$$\bar{V} = \tfrac{1}{4}\mathbf{y}'\mathbf{A}\mathbf{y}. \tag{3.9}$$

In a similar way the mean kinetic energy works out to be

$$\bar{T} = \tfrac{1}{4}k^2 \mathbf{y}'\mathbf{y}. \tag{3.10}$$

Equations (3.9) and (3.10), involving the vector $\mathbf{y}$ and the matrix $\mathbf{A}$, are clearly generalizations of the scalar equations which hold for the one dimensional case. They are of fundamental importance.

We may now ask under what circumstances $\bar{V} = \bar{T}$. If the system is vibrating in a normal mode then $\mathbf{y}$ is an eigenvector and there is a scalar λ such that $\mathbf{A}\mathbf{y} = \lambda \mathbf{y}$. Then

$$\bar{V} = \tfrac{1}{4}\lambda \mathbf{y}'\mathbf{y},$$

and $\bar{V} = \bar{T}$ if $\lambda = k^2$. This is the same conclusion as before; in a normal mode of vibration the square of the angular velocity is equal to an eigenvalue of the matrix $\mathbf{A}$. It is not quite so easy to demonstrate that

$\bar{V} = \bar{T}$ *only if* the system is vibrating in a normal mode. An adaptation of the argument in Section 2.9 will establish the required result, but wishing to stress principles and to reduce detailed calculation as much as possible we will omit a proof.

Further comparison with the geometrical situations in Section 2.9 is appropriate. The really important thing is that for our present purposes any system with n degrees of freedom behaves just as a single particle at the origin in n-dimensional space. When we say 'behaves just as' we mean that it obeys the same algebraic equations. Instead of considering a complicated system we need only consider one particle—at the expense of course of considering it in n-dimensional space. At this stage a warning must be repeated. It is the warning that applies to mathematical models of every kind. The present simplicity has been obtained by making certain assumptions, the most important being those involved in making the linear approximation to the original system. If this ceases to be justified, as it will for example if we wish to study large oscillations in any of the systems under consideration, then our conclusions cease to be valid as physical science, however well they may stand up as pure mathematics.

Looking at the picture of the situation in n-dimensional space, the spheres $\mathbf{y'y}$ = constant are loci of equal kinetic energy and the ellipsoids $\mathbf{y'Ay}$ = constant are loci of equal potential energy. Given the symmetric matrix $\mathbf{A}$ the Rayleigh quotient $\mathbf{y'Ay}/\mathbf{y'y}$ lies between the greatest and least eigenvalues of the matrix. This is Rayleigh's principle, which was discussed in Theorem 2.6. It is quite feasible to obtain estimates of the greatest and least eigenvalues by experimenting with likely vectors, as later detailed examples will show. Obtaining information on intermediate eigenvalues is a little more difficult, but there are ways of doing so.

Exercise 3.3

Normal co-ordinates were mentioned briefly at the end of Section 3.1. Using the ideas of Section 2.8 show that if $\mathbf{H}$ is the matrix whose columns are the normalized eigenvectors of $\mathbf{A}$, and if $\boldsymbol{\eta} = \mathbf{H}\boldsymbol{\xi}$, then the potential energy of the system is

$$\tfrac{1}{2}(\lambda_1 \xi_1^2 + \lambda_2 \xi_2^2 + \dots),$$

and that the kinetic energy is

$$\tfrac{1}{2}(\dot{\xi}_1^2 + \dot{\xi}_2^2 + \dots),$$

where $\lambda_1, \lambda_2, \dots$ are the eigenvalues of $\mathbf{A}$.

This change of co-ordinates facilitates direct algebraic proof of some of the properties now under discussion. For example, we see at once that if we are considering vibrations about a position of stable equilibrium then all of the eigenvalues are positive; because the potential energy has to be positive for any small displacement from a position of stable equilibrium.

EXAMPLE 3.3 (continued)
The matrix associated with the three equally spaced particles of unit mass, on an elastic string is

$$\mathbf{A} = \begin{bmatrix} 2 & -1 & 0 \\ -1 & 2 & -1 \\ 0 & -1 & 2 \end{bmatrix}.$$

It was shown earlier that the smallest eigenvalue associated with the slowest mode of vibration is $\lambda = 2 - \sqrt{2} = 0.586$. If we take as a rather crude approximation to this mode of vibration the trial vector $\mathbf{y} = (1, 1, 1)'$, then

$$\mathbf{y'Ay} = 2 \quad \text{and} \quad \mathbf{y'y} = 3, \quad \text{(verify)},$$

leading to the Rayleigh quotient $\dfrac{2}{3} = 0.667$.

The trial vector $\mathbf{y} = (1, 2, 1)'$ leads to exactly the same result (verify); but $\mathbf{y} = (2, 3, 2)'$ is rather better, leading to a quotient 0.588 which is closer still. A trial vector $\mathbf{y} = (5, 7, 5)'$ gives a quotient which agrees with the true eigenvalue to three figures.

The maximum eigenvalue was shown to be $2 + \sqrt{2} = 3.414$. $(1, -2, 1)'$ gives a Rayleigh quotient of 3.333; $(2, -3, 2)'$ gives 3.412 whereas $(5, -7, 5)'$ gives 3.414.

It must be explained that the vectors chosen as trial vectors are very plausible choices based on a modest degree of physical intuition as to how the system will behave. Also, in all circumstances like this, it is always clear which of two trial vectors is the better one because if the least (greatest) eigenvalue is being sought the better vector is the one giving the smaller (greater) quotient.

Exercise 3.4
The case of n particles equally spaced on a string is treated as Example 4.3 in the next chapter, but the case of four may be studied with profit now. Show that the matrix involved is

$$\begin{bmatrix} 2 & -1 & 0 & 0 \\ -1 & 2 & -1 & 0 \\ 0 & -1 & 2 & -1 \\ 0 & 0 & -1 & 2 \end{bmatrix},$$

which has eigenvalues $2-\gamma$, $3-\gamma$, $1+\gamma$, $2+\gamma$ where γ is the golden ratio number; $\gamma = 1.618$. These numbers are 0.382, 1.382, 2.618, 3.618 respectively. This example is easier to do by the characteristic equation than might have been anticipated because the characteristic equation decomposes into two quadratic factors. The eigenvectors are not difficult to compute and this is left as an exercise. The golden number occurs in each one, and this vibrating system might therefore be called the golden oscillator.

Obtain estimates to $\lambda_{\min}$ and $\lambda_{\max}$ by finding Rayleigh's quotient with suitable trial vectors. ((2, 3, 3, 2)' and (2, −3, 3, −2)' are suggested as a start.)

Find the inverse matrix—a physical argument remembering the interpretation of the terms in **G**, will shorten the work considerably.

Exercise 3.5

Continue with Example 3.4. In this example the least and greatest eigenvalues were stated to be 0·416 and 6·29.

Verify that the trial vectors (4, 2, 1)' and (0, −1, 2)' give quotients of 0·429 and 6·20. Try to find some better trial vectors.

Exercise 3.6

The same problem may be considered with four particles and the reader may show that the relevant matrix is

$$\begin{bmatrix} 1 & -1 & 0 & 0 \\ -1 & 3 & -2 & 0 \\ 0 & -2 & 5 & -3 \\ 0 & 0 & -3 & 7 \end{bmatrix}.$$

(The case of n particles is considered in Example 4.4.) The eigenvalues and eigenvectors, calculated by computer, are

$$\begin{aligned}
\lambda = 0\cdot 323, & \quad (0\cdot 777, 0\cdot 526, 0\cdot 316, 0\cdot 141)'; \\
1\cdot 746, & \quad (0\cdot 598, -0\cdot 446, -0\cdot 579, -0\cdot 330)'; \\
4\cdot 537, & \quad (0\cdot 197, -0\cdot 697, 0\cdot 437, 0\cdot 532)'; \\
9\cdot 395, & \quad (-0\cdot 023, 0\cdot 194, -0\cdot 612, 0\cdot 766)'.
\end{aligned}$$

Investigate approximations to the greatest and least of the eigenvalues by using Rayleigh's principle.

Find the inverse matrix.

The problem now arises of using the Rayleigh quotient with differential operators. It is first necessary to calculate the potential energy of the string in Example 3.4 or the chain in Example 3.5. Using a method which is applicable to any similar system, the potential energy due to the displacement of the element between points x and $x + \delta x$ is

$\frac{1}{2}\eta f \, \delta x +$ higher order terms $= \frac{1}{2}\eta \mathscr{D}\eta \, \delta x +$ higher order terms,

since $f = \mathscr{D}\eta$ by Eq. (3.1).

Proceeding to the limit and integrating over $0 \leqslant x \leqslant 1$, the total potential energy stored when the deflection is $\eta(x)$, is

$$\tfrac{1}{2}\int_0^1 \eta \mathscr{D}\eta \, dx,$$

If $\eta(x, t) = y(x) \sin kt$, working as before, the mean potential energy is

$$\bar{V} = \tfrac{1}{4}\int_0^1 y\mathscr{D}y \, dx. \qquad (3.11)$$

In a similar way, remembering that our systems have unit mass per unit length, the instantaneous kinetic energy is

$$\tfrac{1}{2}\int_0^1 \dot{\eta}^2 \, dx,$$

and the mean kinetic energy during a simple harmonic motion of angular frequency k is, as before,

$$\bar{T} = \tfrac{1}{4}k^2\int_0^1 y^2 \, dx. \qquad (3.12)$$

Equations (3.11) and (3.12) are the analogues for the continuous case of equations (3.9) and (3.10).

All the indications now are to apply Rayleigh's principle to the continuous case, using these two integrals, and hope that it will work! Subsequent examples will show that it is indeed a very good guide. In the continuous cases there are sequences of eigenvalues which tend to infinity, and as there is no greatest eigenvalue the principle cannot be applied at the upper end; but it can still be of assistance in estimating the lowest eigenvalue, and as examples will show, it does in fact do this very satisfactorily. To prove this is another matter, and by current standards Rayleigh himself by no means proved all the ideas which he used in his writing; but he had the foresight to see ahead. Rigorous proof of the ideas now under discussion demands delicate analysis. Our aim is not to provide rigorous proofs, but to indicate what rigorous theories have to establish and to show what the delicate analysis in more advanced books achieves.

EXAMPLE 3.5 (continued)

When Rayleigh's principle is applied to continuous systems trial functions have to be sought in the space of functions which is under consideration—that is the set of functions which

(i) satisfy the boundary conditions of the problem; and
(ii) are functions to which the differential operator may be applied.

In our cases the second condition usually means functions which possess a continuous second derivative throughout the domain involved.

In seeking trial functions one is guided by physical intuition, choosing functions which one expects to be something like the fundamental mode of vibration. In the case of the string problem, if the solution was not already known one might try a function such as

$$y = x(1-x).$$

This satisfies the boundary conditions $y(0) = y(1) = 0$. Remembering that the differential operator for the problem is $\mathscr{D} = -d^2/dx^2$, the Rayleigh quotient is

$$\frac{\int_0^1 y\mathscr{D}y\,dx}{\int_0^1 y^2\,dx} = \frac{1/3}{1/30} = 10.$$

This is an overestimate to the lowest eigenvalue, which exact calculation showed earlier to have $\omega = \pi$ and $\lambda = \omega^2 = \pi^2 = 9{\cdot}87$. A more thorough search, using perhaps higher powers of the expression $x(1-x)$ as well, enables approximations which are smaller, and hence closer to the true value to be found.

EXAMPLE 3.6 (continued)
It is a little more difficult to get close approximations in the case of the swinging chain. It will be remembered that in this example

$$\mathscr{D} = -\frac{d}{dx}\left(x\,\frac{d}{dx}\right),$$

and the lowest eigenvalue was stated to arise with $2\omega = 2{\cdot}405$, which entails $\lambda = \omega^2 = 1{\cdot}442$.

The boundary conditions are $y(0)$ finite and $y(1) = 0$. If we try $y = 1-x$ the resulting quotient turns out to be $1{\cdot}5$. Trying $y = (1-x)^2$ gives $1{\cdot}67$ which is worse, but trying

$$y = (1-x) + (1-x)^2$$

gives $1{\cdot}452$ which is the best so far, and in excess by less than one per cent.

The reader may experiment, trying to do better with other likely functions. Functions of the form $y = 1 - x^\theta$ are relatively easy to manipulate, but the best they will give is a quotient of $1{\cdot}457$ when $\theta = 1/\sqrt{2}$.

Fuller information on the use of Rayleigh's principle, including methods of refining approximations and methods of using it to obtain eigenvalues other than the greatest and least may be found in the works cited in References [1] and [6].

3.6 A geographical application

All of this would seem a far cry from geography, but Gould [3] gives a geographical application of matrices which has ideas in common with the mechanical examples of this chapter.

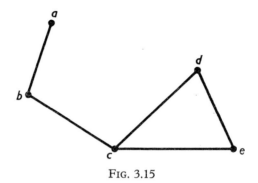

FIG. 3.15

The road network in Fig. 3.15, may be described by the matrix

	a	b	c	d	e
a	1	1	0	0	0
b	1	1	1	0	0
c	0	1	1	1	1
d	0	0	1	1	1
e	0	0	1	1	1

A number of recent school texts introduce connectivity matrices of this kind. The diagonal entries are purely conventional—you may or may not decide to adopt the convention that nodes are connected to themselves. The convention has been adopted here because it has been adopted by geographers in the study of communications networks. The use to which geographers put these matrices is somewhat controversial, but we will indicate how they have been used and leave the reader to make up his own mind.

Two of the eigenvalues of the matrix in the example are relatively easy to find, the others require more elaborate computation. The eigenvalues and normalized eigenvectors are

$$\lambda = 3 \cdot 214, \quad (0 \cdot 155, 0 \cdot 342, 0 \cdot 604, 0 \cdot 497, 0 \cdot 497)';$$
$$2 \cdot 000, \quad (0 \cdot 632, 0 \cdot 632, 0 \cdot 000, -0 \cdot 316, -0 \cdot 316)';$$
$$0 \cdot 461, \quad (0 \cdot 671, -0 \cdot 362, -0 \cdot 476, 0 \cdot 309, 0 \cdot 309)';$$
$$0 \cdot 000, \quad (0 \cdot 000, 0 \cdot 000, 0 \cdot 000, -0 \cdot 707, 0 \cdot 707)';$$
$$-0 \cdot 675, \quad (-0 \cdot 354, 0 \cdot 593, -0 \cdot 639, 0 \cdot 239, 0 \cdot 239)'.$$

The suggestion is that the eigenvectors high-light aspects of the communication system, the eigenvectors associated with the larger eigenvalues being the most important. Thus the first eigenvector attaches the greatest weight to node c, and in some sense seems to pick it out as a favourable communications centre. The next eigenvector looks beyond this and seems to demonstrate in some sort of way the opposite situations of a and b compared with d and e. The interested reader might consider other simple networks in a similar way.

Another line of investigation for the reader is to consider in what ways this treatment resembles the treatment of vibrational problems, and in what ways it differs. Both involve the eigenvectors of symmetric matrices. In the mechanical examples energy considerations bring it about that all of the eigenvalues are positive, this is not so in the geographical applications.

Gould [3] has studied the road network in Uganda using a 18×18 matrix and a 31×31 matrix, and the road network in Syria using a 49×49 matrix. Other geographical applications of eigenvectors are explored in the same reference.

References

1. BICKLEY, W. G. and TALBOT, A., *An Introduction to the Theory of Vibrating Systems*, OUP (1961).
2. EISENSCHITZ, R. K., *Matrix Algebra for Physicists*, Heinemann (1966).
3. GOULD, P. R., 'The geographical interpretation of eigenvalues', *Trans. Inst Br. Geogr.*, **42**, 53–85 (Dec. 1967).
4. HEADING, J., *Matrix Theory for Physicists*, Longmans (1968).
5. PRENTIS, J. M. and LECKIE, F. A., *Mechanical Vibrations*, Longmans (1963).
6. TEMPLE, G. and BICKLEY, W. G., *Rayleigh's Principle*, Reprinted, Dover (1956).

4
Further Examples

This chapter continues the development of the ideas introduced in the previous chapter, presenting some rather more difficult examples based on the same principles. In Section 4.3 difference operators are introduced, and in the present context these are illuminating as they occupy a position between the matrix operators and the differential operators introduced in the last chapter. These examples can be formulated equally well in terms of matrices, but the difference operators enable equations to be constructed whose form is closely analogous to the differential equations describing continuous systems. Some of the later examples are indicated only in outline, and the whole chapter may be omitted at the first reading of the book without loss of continuity.

4.1 The revolving chain

This example concerns the transverse vibrations of a chain which is situated in a plane which rotates with uniform angular velocity Ω about a fixed axis. Science fiction contains many descriptions of space laboratories which are rotated so that the resulting centrifugal force provides artificial gravity. We consider the motion of a chain in such a gravitational field, as it would appear to the astronauts. It turns out that this motion differs from that in Example 3.6 because there the intensity of the gravitational field was constant, whereas here the 'gravity' is proportional to the distance from the axis of rotation.

In Example 3.6 the position co-ordinate on the chain was measured from the free end, whereas here it is measured from the fixed end. The reason for this apparent inconsistency is easily explained. By working in this way we are able to express the solutions in the two cases most simply in terms of known mathematical functions which are fully described and tabulated. In the earlier example Bessel functions were involved; in this example Legendre polynomials appear.

EXAMPLE 4.1

A chain of unit length and unit line density is revolving about an axis perpendicular to its length with constant angular velocity Ω. One end of the chain is fastened to the axis, the other end is free (Fig. 4.1). Considering small transverse displacements in the rotating plane, formulate the equations of motion, and discuss them on the lines of the preceding chapter.

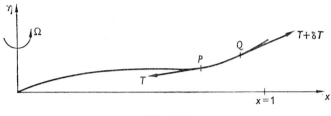

FIG. 4.1

Let $\eta(x)$ be the transverse displacement at the point x of the chain.

If the tension in the chain at P is T, then the component of this tension along the x-axis is $T \cos \theta$, where $\tan \theta = \mathrm{d}\eta/\mathrm{d}x$, and at Q the component is

$$T \cos \theta + \frac{\mathrm{d}}{\mathrm{d}x}(T \cos \theta)\, \delta x + \text{higher order terms.}$$

The difference between these two quantities is the net force on the element PQ. Working to the first order, which means taking $\cos \theta = 1$, after some reduction this comes to $(\mathrm{d}T/\mathrm{d}x)\, \delta x$. The mass of the element is δx, and because of the circular motion the acceleration at P is $\Omega^2 x$ and the acceleration at Q is $\Omega^2(x + \delta x)$, these accelerations being directed towards the axis of rotation through O. Forming the equation of motion and proceeding to the limit

$$\frac{\mathrm{d}T}{\mathrm{d}x} = -\Omega^2 x.$$

Integrating,

$$T = -\tfrac{1}{2}\Omega^2 x^2 + \text{constant.}$$

Since $T = 0$ when $x = 1$ the constant is $\tfrac{1}{2}\Omega^2$.

Hence
$$T = \tfrac{1}{2}\Omega^2(1 - x^2).$$

We now have to consider the transverse motion of the element PQ. As before, the component of the tension at P perpendicular to the x-axis is to the first order

$$T \frac{\mathrm{d}\eta}{\mathrm{d}x}.$$

At Q this component is
$$T\frac{d\eta}{dx} + \frac{d}{dx}\left(T\frac{d\eta}{dx}\right)\delta x + \text{higher order terms}.$$

Taking the difference to get the net force on the element PQ, its equation of motion in the η direction is
$$\ddot\eta\,\delta x = \frac{d}{dx}\left[T\frac{d\eta}{dx}\right]\delta x + \text{higher order terms}.$$

Substituting for T and proceeding to the limit we have
$$\ddot\eta = \frac{d}{dx}\left[\tfrac{1}{2}\Omega^2(1-x^2)\frac{d\eta}{dx}\right].$$

It simplifies the equation if we take a new time variable
$$\tau = \Omega t/\sqrt{2}.$$

In terms of this new variable the equation for transverse motion becomes
$$\ddot\eta = \frac{d}{dx}\left[(1-x^2)\frac{d\eta}{dx}\right],$$

the dot now denoting differentiation with respect to the new time variable τ.

If there is an additional external force $f(x)\,\delta x$ applied to the element PQ, then this equation becomes
$$\ddot\eta = \frac{d}{dx}\left[(1-x^2)\frac{d\eta}{dx}\right] + f(x).$$

If the chain is at rest relative to the rotating frame of reference, then
$$\frac{d}{dx}\left[(1-x^2)\frac{d\eta}{dx}\right] = -f(x).$$

This time putting
$$\mathscr{D} = -\frac{d}{dx}\left[(1-x^2)\frac{d}{dx}\right],$$

the equation takes the usual form
$$\mathscr{D}\eta = f. \tag{4.1}$$

The rest of the investigation follows the now familiar lines. The boundary condition at $x=0$ is $\eta = 0$, and at first sight there is no boundary condition at $x=1$. But the differential equation possesses solutions which tend to infinity as x tends to one, and these clearly do not apply to this problem. Therefore the boundary condition at $x = 1$ is the requirement that the solution remains finite.

Exercise 4.1
  Show that the Green's function for this problem is

$$G(x, \xi) = \tfrac{1}{2} \log \frac{1 + x}{1 - x}, \qquad 0 \leq x \leq \xi,$$

$$= \tfrac{1}{2} \log \frac{1 + \xi}{1 - \xi}, \qquad \xi \leq x \leq 1.$$

Bearing in mind the physical significance of the Green's function, check the plausibility of these equations.

Exercise 4.2
  The eigenvalues and eigenfunctions of the operator $\mathscr{D}$ arise as before from the equation

$$(\mathscr{D} - \omega^2)y = 0, \qquad \text{with } y(0) = 0 \text{ and } y(1) \text{ finite.}$$

Verify that:

x is an eigenfunction with eigenvalue $\omega^2 = 2$
$5x^3 - 3x$ is an eigenfunction with eigenvalue $\omega^2 = 12$
$63x^5 - 70x^3 + 15$ is an eigenfunction with eigenvalue $\omega^2 = 30$.

From this no general rule is apparent. Apart from a numerical constant these functions are the Legendre polynomials of odd order. Legendre polynomials occur in many problems of applied mathematics, especially problems on the potential of spherical distributions of electric charge or gravitation.

The general definition of the nth Legendre polynomial $P_n(x)$ is

$$P_n(x) = \frac{1}{2^n n!} \frac{d^n}{dx^n} \left[(x^2 - 1)^n \right].$$

For odd values of n these provide solutions to the above eigenvalue problem, with eigenvalues $\omega^2 = n(n + 1)$.

It is comparatively easy to verify that these functions provide solutions of the problem, but it is rather harder to show that they are the only solutions. We will assume this.

4.2 Transverse vibrations of beams

The equation governing the transverse vibrations of a beam depends on a number of physical parameters—the density of the beam, the area of cross-section, the moment of inertia of the cross-section, and the modulus of elasticity of the material. We are concerned with the forms which equations take and with the techniques by which they may be solved and not with the numerical values which apply in particular cases, so we will assume that units are adopted which result in the equations appearing in their simplest forms.

EXAMPLE 4.2

It can be shown (see References [8] and [13]) that the transverse displacements of a beam are described by a differential equation of the fourth order,

$$\ddot{\eta} + \eta'''' = 0, \qquad (4.2)$$

where η is the lateral displacement at the point with co-ordinate x, a dot denotes differentiation with respect to t and a prime differentiation with respect to x. Before considering the boundary conditions we will note first an interesting difference between this equation and the equation for the vibrations of a stretched string studied earlier. A simple harmonic oscillation, of angular frequency k, travelling along the x-axis with velocity v is described by the equation

$$\eta = \sin k(t - x/v).$$
$$\ddot{\eta} = -k^2 \sin k(t - x/v)$$

and
$$\eta'''' = (k^4/v^4) \sin k(t - x/v).$$

Hence, if the wave satisfies the differential equation

$$-k^2 + k^4/v^4 = 0,$$

and so
$$v^4 = k^2,$$

or
$$v = \pm\sqrt{k}.$$

This means that the velocity of propagation of the wave depends on its frequency, being higher for waves of high frequency. This is realizable experimentally. A metal pipe satisfies the same equation as a beam, and if a long pipe is hit with a hammer an observer standing some distance along hears the high frequencies before the low ones. With the stretched string it is easy to verify that the velocity of propagation is independent of the frequency. (Do this.)

Now consider the circumstances in which there can be standing waves in the beam, varying sinusoidally with time. These have an equation

$$\eta = y(x) \sin kt.$$

At every point there is a simple harmonic motion of angular frequency k, and $y(x)$ gives the amplitude of the oscillation at the point x. Substituting in Eq. (4.2),

$$y'''' - k^2 y = 0. \qquad (4.3)$$

To have a suitable supply of symbols for later working it is useful to put $k^2 = \mu^4 = \lambda$. The general solution of Eq. (4.3) is then

$$y = A \cos \mu x + B \sin \mu x + C \cosh \mu x + E \sinh \mu x. \qquad (4.4)$$

To get any further with the problem it is necessary to consider the boundary conditions, that is the conditions which hold at the two ends where the beam is supported. We will take these as the points $x = 0$ and $x = 1$.

If the beam is supported freely at an end then the appropriate boundary conditions are

$$y'' = y''' = 0.$$

The engineer will realize that this is saying that the bending moment and the shearing force both vanish.

When the end is clamped the appropriate boundary conditions are

$$y = y' = 0.$$

If the end is pivoted then the appropriate conditions are

$$y = y'' = 0.$$

Other more complicated cases are also possible.

We will consider four typical cases.

Case I

The beam is supported freely at both ends. This means that the modes of vibration are given by the eigenfunctions of the system of equations

$$\begin{aligned} y'''' - \lambda y = 0, & \quad 0 < x < 1; \\ y'' = y''' = 0, & \quad x = 0, x = 1. \end{aligned}$$

The general solution of the first equation is stated in Eq. (4.4) with $\lambda = \mu^4$.

$$y'' = -\mu^2 A \cos \mu x - \mu^2 B \sin \mu x + \mu^2 C \cosh \mu x + \mu^2 E \sinh \mu x.$$

If this vanishes at $x = 0$, then

$$-A + C = 0.$$

If it vanishes at $x = 1$, then

$$-\mu^2 A \cos \mu - \mu^2 B \sin \mu + \mu^2 C \cosh \mu + \mu^2 E \sinh \mu = 0. \qquad (4.5)$$

$$y''' = \mu^3 A \sin \mu x - \mu^3 B \cos \mu x + \mu^3 C \sinh \mu x + \mu^3 E \cosh \mu x.$$

If this vanishes at $x = 0$, then

$$-B + E = 0;$$

and if it vanishes at $x = 1$, then

$$\mu^3 A \sin \mu - \mu^3 B \cos \mu + \mu^3 C \sinh \mu + \mu^3 E \cosh \mu = 0. \quad (4.6)$$

Hence $A = C$ and $B = E$, showing that the eigenfunctions are of the form

$$y = A(\cos \mu x + \cosh \mu x) + B(\sin \mu x + \sinh \mu x);$$

and from Eqs. (4.5) and (4.6)

$$\frac{A}{B} = \frac{\sin \mu - \sinh \mu}{-\cos \mu + \cosh \mu} = \frac{\cos \mu - \cosh \mu}{\sin \mu + \sinh \mu}.$$

After a little reduction, we have

$$\cos \mu \cosh \mu = 1. \quad (4.7)$$

The eigenvalues of the system are given by the roots of this equation, which can be found graphically, or by suitable numerical computation. They were calculated to seven decimal places nearly a century ago by Rayleigh. Apart from $\mu = 0$ the first few roots are

$$\mu = 4{\cdot}730,\ 7{\cdot}853,\ 11{\cdot}00,\ 14{\cdot}14,\ 17{\cdot}28 \dots . \quad (4.8)$$

The larger roots approximate very closely to $(2r + 1)\pi/2$, as can be seen by drawing the graphs of $\cos \mu$ and $1/\cosh \mu$ and seeing where they intersect.

The zero eigenvalue has a physical significance. It shows that a non-oscillatory motion is consistent with the boundary conditions. To recapitulate, the normal modes of vibration are described by expressions of the form

$$\eta = [A(\cos \mu x + \cosh \mu x) + B(\sin \mu x + \sinh \mu x)] \sin kt,$$

where $k = \mu^2$, and they occur for the values of μ given by Eq. (4.8).

Case II

If the beam is clamped at both ends the boundary conditions are

$$y = y' = 0, \quad x = 0, \quad x = 1.$$

Detailed working shows that the same frequencies arise as in the previous case, although $\mu = 0$ is not an eigenvalue this time, and the oscillations are described by

$$\eta = [A(\cos \mu x - \cosh \mu x) + B(\sin \mu x - \sinh \mu x)] \sin kt.$$

Exercise 4.3
Verify this.

These eigenfunctions each involve two arbitrary constants, and therefore in this problem the eigenfunctions associated with each eigenvalue form a vector space of dimension two.

Case III

If the beam is clamped at $x = 0$ and is free at $x = 1$, then the eigenvalues can be shown to be the roots of the equation

$$\cos \mu \cosh \mu = -1,$$

which are

$$\mu = 1{\cdot}875,\ 4{\cdot}694,\ 7{\cdot}855,\ \ldots$$

and thereafter they are indistinguishable from the roots of the previous equation. Note that there is an extra low frequency. The expression for η is as in Case II.

The situation is similar to that with open and closed organ pipes, but with organ pipes the upper frequencies are multiples of the fundamental, whereas with the bar this is not so, and this is why a metal bar vibrating as in a Jews' Harp does not produce a musical note with the qualities of an organ pipe.

Case IV

If the beam is pinned at each end the boundary conditions are

$$y = y'' = 0 \quad \text{when } x = 0 \quad \text{and} \quad x = 1.$$

Adapting the general solution, Eq. (4.4) to these boundary conditions, the eigenfunctions are found to be of the form

$$y = B \sin \mu x \quad \text{where } \mu = r\pi.$$

This case is interesting because the beam vibrates as a string does, and the higher frequencies are integral multiples of the fundamental. However, there is one very important difference. The angular frequencies of the vibrations are given by $k = \mu^2 = r^2\pi^2$. Hence many of the overtones present in the vibrations of a string are absent from the vibrations of a pinned metal strip, and its musical qualities are different.

The eigenvalues and eigenfunctions have other interesting applications. When a beam buckles under a longitudinal stress it can be shown that the critical stress is proportional to the smallest eigenvalue. A proof of this goes a little beyond the techniques we have been using, but the result is plausible as the lateral deformation of the beam takes place under the beam's internal forces, no external transverse forces being applied. Another phenomenon of practical importance is 'whirling' in rotating shafts. When a long, thin shaft is rotated with increasing angular velocity about its own axis a point is reached when it ceases to rotate steadily and 'whirls'. This may be regarded as a form of buckling induced by the centrifugal forces which arise from the rotation. After this there are certain other critical frequencies where whirling occurs. It can be shown that these frequencies are the same as the frequencies of the free transverse vibrations. For more details a textbook on mechanical engineering must be consulted. Hodgson [8] discusses these phenomena clearly and gives many examples.

All of the work done with the vibrating string may be repeated for the vibrating beam. In particular, Green's functions may be found. The appropriate way to define the Green's function has to be decided, however, and this is a good time at which to remember that the Green's function depends on the boundary conditions as well as on the differential operator. With each of the sets of boundary conditions considered above, there is a different Green's function.

The Green's function $G(x, \xi)$ describes the displacement at the point x when a unit force is applied at the point ξ, so it has to satisfy the differential equation

$$y'''' = 0, \quad 0 < x < \xi, \quad \xi < x < 1,$$

and it also has to satisfy the appropriate boundary conditions.

In addition it has to satisfy some condition at the point $x = \xi$, which involves the unit force which is applied: a little investigation shows that the requirements are that y, y' and y'' should be continuous at $x = \xi$, whereas

$$[y''']_{\xi-0}^{\xi+0} = -1.$$

This equation describes a discontinuity in the shearing force at the point ξ.

The general solution of the differential equation is

$$y = Ax^3 + Bx^2 + Cx + E.$$

Fitting the various boundary conditions the Green's functions may be found as follows, the detailed working being left as an exercise:

Case I

When the beam is free at both ends there is no Green's function as the rod would not be stable under the applied unit force. This is directly related to the zero eigenvalue.

Case II

When the beam is clamped at both ends, at $x = 0$, $y = E$ and $y' = C$, so $E = C = 0$. Hence for $0 \leq x \leq \xi$ the Green's function is of the form $y = A_1 x^3 + B_1 x^2$. For $\xi \leq x \leq 1$ it is similarly of the form $y = A_2(1-x)^3 + B_2(1-x)^2$.

If y, y' and y'' are continuous at $x = \xi$, and there is the discontinuity of y''' of the stated magnitude, then, after some calculation we find

$$y = \tfrac{1}{6}x^2(1-\xi)^2(2x\xi + x - 3\xi), \quad 0 \leqslant x \leqslant \xi;$$
$$= \tfrac{1}{6}\xi^2(1-x)^2(2\xi x + \xi - 3x), \quad \xi \leqslant x \leqslant 1;$$

and this is the required Green's function $G(x, \xi)$.

Case III

When the beam is clamped at $x = 0$, and free at $x = 1$, the Green's function is

$$G(x, \xi) = \tfrac{1}{6}x^3 - \tfrac{1}{2}\xi x^2, \quad 0 \leqslant x \leqslant \xi;$$
$$= \tfrac{1}{6}\xi^3 - \tfrac{1}{2}\xi^2 x, \quad \xi \leqslant x \leqslant 1.$$

Case IV

When each end of the beam is pinned the Green's function is

$$G(x, \xi) = \tfrac{1}{6}x(1-\xi)(x^2 + \xi^2 - 2\xi), \quad 0 \leqslant x \leqslant \xi;$$
$$= \tfrac{1}{6}\xi(1-x)(\xi^2 + x^2 - 2x), \quad \xi \leqslant x \leqslant 1.$$

Exercise 4.4

Sketch a number of the above functions, and check that they agree with physical intuition.

As before, arguing physically, it is plausible that if a unit force applied at the point $x = \xi$ produces the deflection

$$\eta = G(x, \xi),$$

then a force distributed along the beam, with $f(\xi)\,\delta\xi$ acting on the interval $(\xi, \xi + \delta\xi)$, will produce a deflection

$$\eta = \int_0^1 G(x, \xi) f(\xi)\, d\xi. \tag{4.9}$$

Working as in the previous chapter $\eta(x)$ and $f(x)$ are also related by the differential operator $\mathscr{D} = d^4/dx^4$, with

$$\mathscr{D}\eta = f.$$

Writing Eq. (4.9) in the form

$$\eta = \mathscr{G}f,$$

using $\mathscr{G}$ to denote the operator on the right of the equation, we once again have a pair of inverse operators $\mathscr{D}$ and $\mathscr{G}$. These are interesting because $\mathscr{D}$ involves four differentiations whereas $\mathscr{G}$ involves only one integration. This underlines the point in the previous chapter where the differential operators were of the second order and the inverse integral operators involved one integration. This is a warning against the too facile assumption that integration and differentiation are the inverses of one another in all circumstances. These problems involve the boundary conditions also. The boundary conditions are very much part of the physical problem, and in the mathematical problem, as the last example shows, they affect the Green's function very much indeed.

The vibrating beam also provides good examples on the application of Rayleigh's principle. In each of Cases II–IV we will seek to obtain over-estimates of the lowest eigenvalue by using trial functions. It will be remembered that trial functions have to satisfy the boundary conditions of the problem, and it has to be possible to apply the differential operator, in this case $\mathscr{D} = \mathrm{d}^4/\mathrm{d}x^4$, to them. Suitable trial functions are polynomials of degree four which satisfy the boundary conditions. In each case the reader should sketch the curves involved and see that they are intuitively plausible approximations to the fundamental mode of oscillation of the beam.

Case II

The beam is clamped at both ends, and satisfies the boundary conditions
$$y = y' = 0, \quad x = 0, \quad x = 1.$$
The polynomial of degree four satisfying these conditions is
$$y = x^2(1 - x)^2.$$
The Rayleigh quotient, omitting details of the working, is
$$\int_0^1 y\, y'''' \, \mathrm{d}y \bigg/ \int_0^1 y^2 \, \mathrm{d}y = \frac{4/5}{1/630} = 504.$$
We saw before that $\mu_{\min} = 4 \cdot 730$; so $\lambda_{\min} = \mu^4_{\min} = 500$, and the approximation has an error of less than one per cent.

Case III

With the boundary conditions
$$y(0) = y'(0) = y''(1) = y'''(1) = 0$$
a suitable trial function is
$$y = 6x^2 - 4x^3 + x^4.$$
With this function the Rayleigh quotient works out to be 12·46. We saw earlier that $\mu_{\min} = 1 \cdot 875$; so $\lambda_{\min} = \mu^4_{\min} = 12 \cdot 36$, and once again we have an approximation to within one per cent.

Case IV

With the boundary conditions
$$y(0) = y''(0) = y(1) = y''(1) = 0$$

a suitable trial function is

$$y = x^4 - 2x^3 + x.$$

This gives a Rayleigh quotient of 97·55, whereas the true eigenvalue is $\lambda_{\min} = \mu^4_{\min} = \pi^4 = 97\cdot42$.

Case I has not been considered; but in this case there is a zero eigenvalue, and merely by taking $y = $ constant we have the first eigenfunction and the Rayleigh quotient is zero.

4.3 The use of difference equations

Vibrational problems may also be studied with the aid of sequences and difference operators. These ideas we will shortly explain, but our strategic interest in them comes from the fact that they indicate in a very clear way the transition from the discrete cases, discussed with matrices, to the continuous cases, discussed with differential equations. In all of the following examples the equations may be formulated just as well in matrix terms, but using difference operators draws attention to the connexions with the continuous problems, producing equations of very similar form, but equations which have to be solved by algebraic techniques instead of by calculus.

Sets of infinite sequences provide examples of interesting vector spaces which are, loosely speaking, somewhere between the spaces of finite dimension involved in the problems with a finite number of particles and the spaces of continuous functions involved in the continuous problems; but in the examples to follow the sequences are always finite, although the number of terms involved can be arbitrarily large. This is because all of our problems are concerned with vibrations in bounded regions—in topological terminology our spaces are *compact* —and over a bounded region we can approximate to continuous functions by taking a sufficiently large, but finite, number of 'sample' points. Thus for numerical work over a bounded two-dimensional region we may approximate by filling the region with a square mesh and considering the values of the functions involved at the mesh points only.

We will start by considering an extension of Example 3.3, with N particles instead of three.

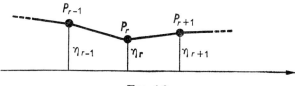

FIG. 4.2

EXAMPLE 4.3

N particles of unit mass are spaced at unit intervals on an elastic string which is fastened to rigid supports at the two ends (Fig. 4.2). Investigate the normal modes of transverse vibration.

[It may be appropriate to remark that in systems such as this the mathematical equations governing longitudinal vibrations are, to a first order approximation, of the same form as those governing transverse vibrations; but we will not prove this.]

Working just as with Example 3.3, taking the tension in the string as unity and dealing with the case when there are no externally imposed forces, the equation giving the net force in the direction of η_r increasing is to the first order

$$(\eta_{r+1} - \eta_r) + (\eta_{r-1} - \eta_r);$$

and so the equation of motion of the rth particle is

$$\eta_{r+1} - 2\eta_r + \eta_{r-1} = \ddot{\eta}_r, \qquad r = 1, \ldots, N.$$

The equations for particles 1 and N are slightly different, but they may be put in this form if we adopt the conventions

$$\eta_0 = \eta_{N+1} = 0.$$

These are the boundary conditions on the sequence.

If there are free oscillations such that

$$\boldsymbol{\eta} = \mathbf{y} \sin \omega t$$

the equations become

$$y_{r+1} - 2y_r + y_{r-1} = -\omega^2 y_r, \qquad r = 1, \ldots, N;$$

or

$$-y_{r+1} + (2 - \omega^2) y_r - y_{r-1} = 0, \qquad r = 1, \ldots, N. \qquad (4.10)$$

Using matrices for the moment, these may be written

$$(\mathbf{A} - \omega^2 \mathbf{I})\mathbf{y} = 0,$$

where

$$\mathbf{A} = \begin{bmatrix} 2 & -1 & 0 & . & . & 0 \\ -1 & 2 & -1 & . & . & 0 \\ 0 & -1 & 2 & . & . & 0 \\ . & & & . & & . \\ 0 & 0 & 0 & . & . & 2 \end{bmatrix}. \qquad (4.11)$$

The matrix $\mathbf{A}$ has N rows and columns and the only non-zero terms are situated on the principal diagonal and on the diagonals immediately above and below. Such a matrix may be called a *banded matrix*.

Starting with the identity
$$\sin(r+1)\theta + \sin(r-1)\theta = 2\cos\theta \sin r\theta,$$
and rearranging
$$-\sin(r-1)\theta + 2\sin r\theta - \sin(r+1)\theta = 2(1-\cos\theta)\sin r\theta. \quad (4.12)$$

If we put $\lambda = 2(1-\cos\theta)$, we see that Eq. (4.12) indicates that λ is an eigenvalue of **A**, with eigenvector $(\sin\theta, \sin 2\theta, \ldots, \sin N\theta)'$, provided that the boundary condition $y_{N+1} = \sin(N+1)\theta = 0$ is satisfied. ($y_0 = \sin 0 = 0$ is satisfied immediately.)

This requires that
$$\sin(N+1)\theta = 0,$$
so
$$\theta = \frac{r\pi}{N+1}, \quad r = 1, \ldots, N.$$

Hence the eigenvalues of **A** are given by
$$\omega_r^2 = \lambda_r = 2\left(1 - \cos\frac{r\pi}{N+1}\right), \quad r = 1, \ldots, N; \quad (4.13)$$

which means that the natural angular frequencies are given by
$$\omega_r = 2\sin\frac{r}{N+1}\cdot\frac{\pi}{2}, \quad r = 1, \ldots, N.$$

The rth eigenvector is
$$\mathbf{y} = \left(\sin\frac{r\pi}{N+1}, \sin\frac{2r\pi}{N+1}, \sin\frac{3r\pi}{N+1}, \ldots\right)'. \quad (4.14)$$

By letting N tend to infinity, and taking suitable care, it is possible to deduce from this all of the results in Example 3.5 for the continuous string (see Reference [13]).

The finite difference operator δ, called the *central difference operator*, may be used to operate on a sequence $s_1, s_2, s_3, \ldots$ and it is defined by the following relation
$$\delta s_r = s_{r+\frac{1}{2}} - s_{r-\frac{1}{2}}.$$

(This symbol δ is *not*, of course, the same symbol as that used in incremental notation in the calculus, as in $x + \delta x$.)

The interest is frequently in cases where the sequence is obtained as sample values of a continuous function, taken at integral values of the variable, and then a meaning can be given to expressions such as $s_{r+\frac{1}{2}}$. However, the operator may be used in other circumstances, as here, because
$$\begin{aligned}\delta^2 s_r = \delta(\delta s_r) &= \delta(s_{r+\frac{1}{2}}) - \delta(s_{r-\frac{1}{2}}), \\ &= (s_{r+1} - s_r) - (s_r - s_{r-1}), \\ &= s_{r+1} - 2s_r + s_{r-1}.\end{aligned}$$

Hence Eq. (4.10) can be written

$$-\delta^2 y_r - \omega^2 y_r = 0.$$

Writing $\mathscr{D}$ to denote now $-\delta^2$ this equation is of exactly the same form as the differential equation in Example 3.5.

To summarize, we may consider the set of sequences of real numbers which satisfy the boundary conditions $y_0 = y_{N+1} = 0$; these form a vector space. We may ask what are the eigenvalues and eigensequences of the operator $\mathscr{D} = -\delta^2$. The previous working provides the answers; the eigenvalues are given by Eq. (4.13) and the eigensequences by Eq. (4.14).

Working the way we have the difference equations have arisen just as the matrix equations did earlier, that is by considering a finite number of particles. Mathematicians sometimes reach this same destination from a different starting point. It is possible to start from the differential equation and make a discrete approximation. This will be discussed in Chapter 5.

Using the difference operator notation, we will now find what we might call the *Green's sequence* for the problem. This is defined in the usual way.

Impose a unit force on the particle r and determine the resulting displacement of every particle. We require the sequence $\eta_0, \eta_1, \eta_2, \ldots, \eta_{N+1}$ which satisfies the equations

$$\delta^2 \eta_n = 0, \quad 0 < n < r, \quad r < n < N+1;$$
$$\eta_0 = \eta_{N+1} = 0,$$

and the extra condition at the 'kink', which a little consideration shows is

$$\delta^2 \eta_r = (\eta_{r+1} - \eta_r) - (\eta_r - \eta_{r-1}) = -1; \quad (4.15)$$

this equation arising because the tensions in the two adjacent segments of the string have to counterbalance the applied unit force.

The sequences which satisfy the condition

$$\delta^2 \eta_n = \eta_{n+1} - 2\eta_n + \eta_{n-1} = 0$$

are arithmetical progressions. For the left-hand portion of the string the solution required is of the form

$$\eta_n = an,$$

(incorporating the condition $\eta_0 = 0$), and for the right-hand portion of the string it is of the form

$$\eta_n = b(N+1-n),$$

(putting it in this form in order to incorporate the condition $\eta_{N+1} = 0$ as simply as possible).

These have to match at $n = r$, giving
$$ar = b(N + 1 - r),$$
and also satisfy Eq. (4.15) which requires, after a little reduction,
$$a + b = 1.$$
Hence the required sequence is specified by
$$\begin{aligned}\eta_n &= (N + 1 - r)n/(N + 1), & 0 \leqslant n \leqslant r, \\ &= (N + 1 - n)r/(N + 1), & r \leqslant n \leqslant N + 1.\end{aligned}$$
This gives the displacement of the particles in Fig. 4.3, and to conform

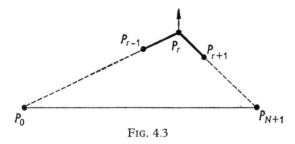

Fig. 4.3

with the notation used for differential equations the terms of this 'Green's sequence' might be denoted by $G(n, r)$. If these sequences are tabulated in a matrix we obtain

$$\begin{array}{c} \\ n = 1 \\ n = 2 \\ n = 3 \\ n = 4 \\ \\ \\ n = N \end{array} \begin{array}{cccccc} r=1 & r=2 & r=3 & r=4 & & r=N \\ \left[\begin{array}{cccccc} N & N-1 & N-2 & N-3 & . & 1 \\ N-1 & 2(N-1) & 2(N-2) & 2(N-3) & . & 2 \\ N-2 & 2(N-2) & 3(N-2) & 3(N-3) & . & 3 \\ N-3 & 2(N-3) & 3(N-3) & 4(N-3) & . & 4 \\ . & . & . & . & . & . \\ . & . & . & . & . & . \\ 1 & 2 & 3 & 4 & . & N \end{array}\right] \div (N+1). \end{array}$$

This matrix is the inverse of the matrix **A** in Eq. (4.11).

Exercise 4.5
Check that this matrix is the inverse of **A**.

Having obtained the Green's function in the present context of sequences and difference operators we might observe that it continues to have all the familiar properties. For example, given the problem of finding the displacements η which result from a set of imposed static forces **f**, we have to solve
$$\mathscr{D}\eta = \mathbf{f}, \tag{4.16}$$

which was the familiar Eq. (3.1) of Chapter 3. $\mathscr{D}$ now denotes $-\delta^2$, and the previous boundary conditions apply. To emphasize the sequence point of view the solution may be written

$$\eta_n = \sum_{r=1}^{N} G(n, r) f_r. \qquad (4.17)$$

This gives the displacement at the point n when a set of forces $(f_1, f_2, \ldots, f_N)$ is applied at points $1, 2, \ldots, N$. Some writers write matrix multiplications in this form as their normal matrix notation, although we have not done so in this book. On the other hand, replacing the summation sign by an integral sign, as we might hope to do if we could justify the limiting process, we recover the familiar equation of the continuous problems in Chapter 3,

$$\eta(x) = \int G(x, \xi) f(\xi) \, d\xi.$$

This shows very clearly how well the sequence point of view relates the discrete and the continuous problems.

Exercise 4.6

If you are in any doubt, verify that Eq. (4.17) is the solution of Eq. (4.16), with the appropriate boundary conditions.

In a similar way we will now discuss the extension of Example 3.4 to N particles.

EXAMPLE 4.4

N equally spaced particles of unit mass hang vertically on a light inextensible string. Investigate the motion for small transverse oscillations in a vertical plane, and determine the Green's function.

As in Example 3.4 we will number the particles from the lowest particle at the free end of the string, and we will continue with the previous notation.

We have, as before, that the tension in the segment of string joining the nth and $(n+1)$th particles is

$$T_n = n.$$

The net transverse force on the nth particle, arising from the horizontal components of the tensions in the strings above and below it, is

$$T_{n-1}(\eta_{n-1} - \eta_n) - T_n(\eta_n - \eta_{n+1}). \qquad (4.18)$$

This applies for $n = 1, \ldots, N$, and if we adopt the conventions that

$$T_0 = 0 \quad \text{and} \quad \eta_{N+1} = 0$$

these expressions become

$$(n-1)\eta_{n-1} - (2n-1)\eta_n + n\eta_{n+1} = \ddot{\eta}_n, \qquad n = 1, \ldots, N;$$

and the stiffness matrix is therefore

$$\mathbf{A} = \begin{bmatrix} 1 & -1 & 0 & 0 & \cdot & \cdot & & \cdot \\ -1 & 3 & -2 & 0 & \cdot & \cdot & & \cdot \\ 0 & -2 & 5 & -3 & \cdot & \cdot & & \cdot \\ 0 & 0 & -3 & 7 & \cdot & \cdot & & \cdot \\ \cdot & \cdot & \cdot & \cdot & \cdot & \cdot & & -(N-1) \\ \cdot & \cdot & \cdot & \cdot & \cdot & -(N-1) & & 2N-1 \end{bmatrix}.$$

The eigenvalues and eigenfunctions of $\mathbf{A}$ determine the normal frequencies and the normal modes. It does not seem possible to give an explicit expression for these, and evaluating them for various values of N is a matter for numerical analysis on a computer. Using the central difference operator δ the expression for the net transverse force can be written as

$$-T_{n-1}\,\delta\eta_{n-\frac{1}{2}} + T_n\,\delta\eta_{n+\frac{1}{2}} = \delta(T_{n-\frac{1}{2}}\,\delta\eta_n).$$

This holds whatever the function giving T in terms of n, but in this case $T_n = n$, and so the net force on particle n arising from the segments of string joining it to its neighbours is $\delta[(n-\frac{1}{2})\,\delta\eta_n]$.

When finding the Green's function a unit external force is imposed on particle r, and the other particles take up equilibrium positions under their mutual interactions and so we have

$$\delta[(n-\tfrac{1}{2})\,\delta\eta_n] = 0, \qquad n \neq r; \\ = -1, \qquad n = r. \tag{4.19}$$

(The reader may need to check the sign carefully once more.)

To calculate the Green's function we may proceed as follows. Working from Eq. (4.19) with the difference operator to illustrate its use, although we could just as easily work directly from the expression (4.18),

$$n\,\delta\eta_{n+\frac{1}{2}} - (n-1)\,\delta\eta_{n-\frac{1}{2}} = 0.$$

When $n < r$ we have

$$n\delta\eta_{n+\frac{1}{2}} = (n-1)\,\delta_{n-\frac{1}{2}} = (n-2)\,\delta_{n-\frac{3}{2}} = \ldots = 1.\delta_{\frac{3}{2}} = 0.\delta_{\frac{1}{2}} = 0.$$

Hence for these values of n

$$\delta\eta_{n+\frac{1}{2}} = 0,$$

which means that

$$\eta_{n+1} - \eta_n = 0$$

or

$$\eta_n = \text{constant}.$$

This of course we knew without any working if we kept in mind the physical interpretation of the Green's function. (If in doubt refer back to Fig. 3.13.) For the upper part of the string the working is more difficult. When $r < n \leq N$, remembering that $\eta_{N+1} = 0$, we have as before that $n\,\delta\eta_{n+1/2}$ is constant, but this time the constant is not zero. This time put

$$n\,\delta\eta_{n+\frac{1}{2}} = n(\eta_{n+1} - \eta_n) = c.$$

Then
$$\eta_{N+1} - \eta_N = \frac{c}{N}$$

so
$$\eta_N = -\frac{c}{N}.$$

Also
$$\eta_N - \eta_{N-1} = \frac{c}{N-1},$$

and
$$\eta_{N-1} = -c\left(\frac{1}{N} + \frac{1}{N-1}\right).$$

Similarly,
$$\eta_{N-s} = -c\left(\frac{1}{N} + \frac{1}{N-1} + \cdots + \frac{1}{N-s}\right).$$

At $n = r$ we have

$$r(\eta_{r+1} - \eta_r) - (r-1)(\eta_r - \eta_{r-1}) = -1;$$

but $\eta_{r-1} = \eta_r$ and so

$$\eta_{r+1} - \eta_r = -\frac{1}{r}.$$

Thus $c = -1$, and after a little manipulation the Green's function is given by

$$G(n, r) = \frac{1}{r} + \frac{1}{r+1} + \frac{1}{r+2} + \cdots + \frac{1}{N}, \quad 1 \leq n \leq r;$$

$$= \frac{1}{n} + \frac{1}{n+1} + \frac{1}{n+2} + \cdots + \frac{1}{N}, \quad r \leq n \leq N.$$

It is worth writing out the Green's matrix at length, and this is made easier if we put

$$g_r = \frac{1}{r} + \frac{1}{r+1} + \cdots + \frac{1}{N}.$$

The Green's matrix is then

$$\mathbf{G} = \begin{bmatrix} g_1 & g_2 & g_3 & \cdot & \cdot & g_N \\ g_2 & g_2 & g_3 & \cdot & \cdot & g_N \\ g_3 & g_3 & g_3 & \cdot & \cdot & g_N \\ \cdot & \cdot & \cdot & \cdot & & \cdot \\ g_N & g_N & g_N & \cdot & \cdot & g_N \end{bmatrix}.$$

The next example is the discrete analogue of Example 4.1. The working will only be indicated in outline.

EXAMPLE 4.5

With the conditions of Example 4.1 consider a light string with N particles of unit mass spaced at equal intervals along it (Fig. 4.4). Investigate the transverse motion and find the Green's function.

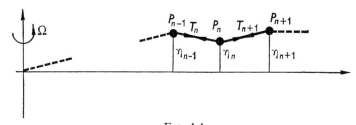

FIG. 4.4

The first step is to find the tension in each segment of the string. Working with the usual first order approximations, and noting that the acceleration of the nth particle towards 0 is $\Omega^2 n$, we have

$$T_{n+1} - T_n = -\Omega^2 n;$$

this equation holding for $n = 1$ up to N, the last equation being

$$T_{N+1} - T_N = -\Omega^2 N$$

where we define T_{N+1} to be zero. Adding the $N - n + 1$ equations, from n to N, and using $T_{N+1} = 0$, we have

$$-T_n = -\Omega^2[n + (n+1) + (n+2) + \ldots + N].$$

So $\quad T_n = \tfrac{1}{2}\Omega^2(N + n)(N - n + 1).$

The net transverse force on the nth particle, from its neighbouring segments of string, is

$$T_{n+1}(\eta_{n+1} - \eta_n) + T_n(\eta_{n-1} - \eta_n); \quad (\eta_0 = 0)$$

or rearranging,

$$T_n \eta_{n-1} - (T_n + T_{n+1})\eta_n + T_{n+1}\eta_{n+1}.$$

As before, with a suitable time variable $\tfrac{1}{2}\Omega^2$ may be put equal to unity. The stiffness matrix is then

$$\mathbf{A} = \begin{bmatrix} T_1 + T_2 & -T_2 & 0 & 0 & . & . \\ -T_2 & T_2 + T_3 & -T_3 & 0 & . & . \\ 0 & -T_3 & T_3 + T_4 & -T_4 & . & . \\ 0 & 0 & -T_4 & T_4 + T_5 & . & . \\ . & . & . & . & . & -T_N \\ . & . & . & . & -T_N & T_N \end{bmatrix},$$

where $T_n = (N + n)(N - n + 1)$.

Exercise 4.7

Write out this matrix for the cases $N = 3$ and $N = 4$, and find the eigenvalues and eigenvectors in these cases. It is to be expected that the calculations would be very tedious, but in these examples the numbers are unusually co-operative. It turns out that the eigenvalues are the first few eigenvalues of the continuous case, Example 4.1. Thus when $N = 3$ the eigenvalues are 2, 12, 30 and when $N = 4$ they are 2, 12, 30, 56. Verify this and find the eigenvectors.

To evaluate the Green's function we impose a unit force on the rth particle and we have to solve the equations

$$T_{n+1}(\eta_{n+1} - \eta_n) - T_n(\eta_n - \eta_{n-1}) = 0, \quad n \neq r,$$
$$= -1, \quad n = r;$$

with $T_{N+1} = 0$ and $\eta_0 = 0$ as boundary conditions. We easily see (if only by ready appreciation of the physics) that for $r < n \leqslant N$, $\eta_n = \eta_r$. Whereas for

$$1 \leqslant n \leqslant r, \quad T_n(\eta_n - \eta_{n-1}) = \text{constant}.$$

Using the condition when $n = r$ we see that the constant is unity. We then have in turn

$$\eta_1 - \eta_0 = 1/T_1,$$
$$\eta_2 - \eta_1 = 1/T_2, \text{ etc.}$$

up to

$$\eta_r - \eta_{r-1} = 1/T_r.$$

Since $\eta_0 = 0$, adding the first n equations,

$$\eta_n = 1/T_1 + 1/T_2 + \ldots + 1/T_n, \quad 1 \leqslant n \leqslant r.$$

Therefore the Green's function may be written

$$G(n, r) = 1/T_1 + 1/T_2 + \ldots + 1/T_n, \quad 1 \leqslant n \leqslant r,$$
$$= 1/T_1 + 1/T_2 + \ldots + 1/T_r, \quad r \leqslant n \leqslant N;$$

with $T_n = (N + n)(N - n + 1)$.

Exercise 4.8

Evaluate the flexibility matrices **G** for the cases $N = 3$ and $N = 4$, and verify that they are the inverses of the stiffness matrices.

Fort [6] may be consulted for more details on the use of difference equations in problems of the type discussed in this section.

4.4 Circulants

Various examples have shown that differences in the boundary conditions can change the problem considerably, even though the differential equation governing the motion remains the same.

It might be expected that when the number of particles in a chain is large the boundary conditions are of relatively less importance in determining what is happening away from the ends of the chain. Therefore, when investigating the vibrations of such a system, it is feasible to make a first attack on the problem by solving it with boundary conditions which are easy to handle. As an illustration we will continue the discussion of Example 4.3. In this we may imagine particle 1 attached to another particle vibrating in phase with particle N, and particle N attached to another particle in phase with particle 1. This is exactly the same mathematically as dealing with particles arranged in a circle.

The effect of this is to modify the stiffness matrix **A**, Eq. (4.11), by putting an additional element -1 in the positions $(1, N)$ and $(N, 1)$. Call this new matrix **B**, so that

$$\mathbf{B} = \begin{bmatrix} 2 & -1 & 0 & . & . & -1 \\ -1 & 2 & -1 & . & . & 0 \\ 0 & -1 & 2 & . & . & 0 \\ . & . & . & . & . & . \\ -1 & 0 & 0 & . & . & 2 \end{bmatrix}. \quad (4.20)$$

This matrix has the property that all of the rows are successive cyclic permutations of the first row. Such a matrix is sometimes called a *circulant*. It is possible to make simple, general statements about the eigenvalues and eigenvectors of circulants and this makes the solution of the related vibrational problems correspondingly easy.

Consider the general circulant, of order n,

$$\mathbf{C} = \begin{bmatrix} a & b & c & . & . & f & g \\ g & a & b & . & . & . & f \\ f & g & a & . & . & . & . \\ . & . & . & . & . & . & . \\ b & c & . & . & . & g & a \end{bmatrix}.$$

Multiplying the column vector $(1, \theta, \theta^2, \ldots, \theta^{n-1})'$, where θ is an nth root of unity, by **C** we get the column vector

$$\begin{bmatrix} a + b\theta + c\theta^2 + . & . + g\theta^{n-1} \\ a\theta + b\theta^2 + c\theta^3 + . & . + g \\ a\theta^2 + b\theta^3 + c\theta^4 + . & . + g\theta \\ . & . \\ a\theta^{n-1} + b + c\theta + . & . + g\theta^{n-2} \end{bmatrix}.$$

This shows that $(1, \theta, \theta^2, \ldots, \theta^{n-1})'$ is an eigenvector with the eigenvalue

$$\lambda = a + b\theta + c\theta^2 + \ldots + g\theta^{n-1}.$$

Taking the n nth roots of unity gives the n eigenvalues and eigenvectors.

Applying this to the matrix **B**, which is of order N, the eigenvalues are

$$\lambda_r = 2 - (\theta_r + \theta_r^{N-1}), \qquad \theta_r = e^{2\pi i r/N},$$

$$= 2 - 2\cos\frac{2\pi r}{N},$$

$$\lambda_r = 4\sin^2\frac{\pi r}{N}, \qquad r = 1, \dots, N.$$

This means that the natural angular frequencies of vibration are

$$\omega_r = 2\sin\frac{\pi r}{N}, \qquad r = 1, \dots, N.$$

One of the eigenvalues is zero. The physical significance of this is that non-oscillatory motion is now possible. A uniform translation of the circular chain normal to its plane is consistent with the equations of motion. **B** has no inverse, and this is all part of the same picture.

There are further slight complications which it will not be necessary to pursue. In this system the eigenvalues coincide in pairs since

$$\sin\frac{\pi r}{N} = \sin\frac{\pi(N-r)}{N}.$$

The system is *degenerate* and the number of frequencies is not equal to the number of degrees of freedom. This means, incidentally, that real eigenvectors can be found as well as the eigenvectors $(1, \theta, \dots, \theta^{N-1})'$.

On an atomic scale the great part of the information which can be found is spectroscopic, that is to say it is information about vibrations; and so work of this kind is basic in the theory of vibrations in molecules and in crystal lattices. An extensive review of the applications to atomic vibrations in solids may be found in a paper by Dean [3]. Of course, there the problems are three-dimensional but the principles are the same.

4.5 Vibrations in cable nets

The same mathematical ideas occur, but on a very different physical scale, in the analysis of the vibrations of suspension bridges and of the networks of cables which are the main load-carrying elements in certain designs of large-span roofs. Laboratory experiments show that it is a reasonable approximation to regard these cable nets as being composed of elastic links joining masses which are concentrated at their points of intersection, that is to say the systems are the two-dimensional analogues of the chains which have just been described. The results which can be obtained are a very plausible two-dimensional extension of the results for vibrating chains, and they will only be summarized here. Proofs may be found in a paper by Buchholdt, Davies and Hussey [2].

Consider a horizontal lattice of points (x, y), with x and y integral, $0 \leqslant x \leqslant M + 1$, $0 \leqslant y \leqslant N + 1$. The boundary points, with $x = 0$, $y = 0$, $x = M + 1$, or $y = N + 1$, are fixed and the MN points inside the rectangle are free to vibrate.

We need to define the central difference operators δ_m and δ_n which operate in the x and y directions. Then giving unit values to certain physical parameters in order to concentrate on the mathematical form of the solution, if z_{mn} is the vertical displacement at the point (m, n) and f_{mn} the imposed force, the equations for the system are

$$(\delta_m^2 + \delta_n^2) z_{mn} = f_{mn}, \quad \text{with } z_{mn} = 0 \text{ on the boundary.} \quad (4.21)$$

For this system the standard questions may be asked—what is the inverse operator? what is the Green's function? what are the normal modes of vibration? and what is the response to imposed periodic forces?

When finding the eigenvalues of the operator on the left it corresponds most closely with the notation employed previously if it is taken with a negative sign. The eigenvalues are then

$$\lambda_{rs} = 2\left(1 - \cos\frac{r\pi}{M+1}\right) + 2\left(1 - \cos\frac{s\pi}{N+1}\right), \quad \begin{array}{l} r = 1, \ldots, M, \\ s = 1, \ldots, N. \end{array}$$

The square roots of these quantities give the angular frequencies of the normal modes of vibration.

The eigenfunctions corresponding, giving the configurations of the network as it oscillates in normal modes, are

$$z_{rs} = \sin\frac{pr\pi}{M+1} \sin\frac{qs\pi}{N+1}, \quad \begin{array}{l} p = 1, \ldots, M, \\ q = 1, \ldots, N. \end{array}$$

This gives the displacement of the point (p, q) in the mode (r, s).

These results apply to the simplest case, which is a lattice of squares with sides parallel to the surrounding rectangular frame; but the method applies to other forms of lattice as well, and Reference [2] gives some interesting examples with the operators and the eigenvalues which correspond.

4.6 Oscillations in electrical filters

Consider the electrical filter shown in Fig. 4.5.

This consists of inductances in series with capacitors in parallel. If there are N stages of the kind shown, and all the inductances have value L, and all the capacitors value C, then, denoting the voltage across the rth capacitor by v_r, the current into it by i_r and the current in the rth inductance by j_r, electrical theory shows that the circuit equations are

$$C\dot{v}_r = i_r,$$
$$j_r = j_{r-1} - i_{r-1},$$
$$L\dot{j}_r = v_{r-1} - v_r.$$

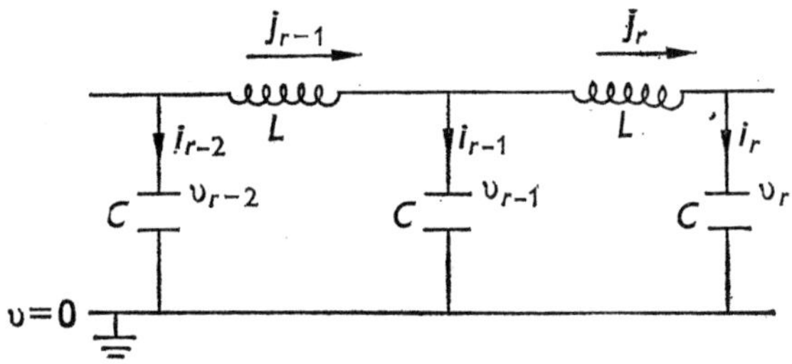

Fig. 4.5

From this we may deduce

$$L\ddot{j}_r = \dot{v}_{r-1} - \dot{v}_r = (i_{r-1} - i_r)/C,$$

and so
$$LC\ddot{j}_r = j_{r-1} - 2j_r + j_{r+1}.$$

These equations will be accompanied by appropriate conditions at the boundaries. The main point of interest is that the equations are of precisely the same form as those for the particles vibrating on a chain in Section 4.3. This means that the previous mechanical discussion applies also to this electrical problem. Furthermore, just as the continuous vibrating string may be regarded as the limiting case of a light string loaded with particles, so a transmission line is the continuous analogue of a filter made up of lumped components. The differential equations of the transmission lines and vibrating strings are of the same mathematical form, although one system is electrical and the other is mechanical.

Further details concerning the electrical applications of matrices may be found in Tropper [12]. An interesting alternative approach to filters is to use 2 × 2 transfer matrices; these are explained also in References [1] and [5]. For mechanical applications, in particular to the study of oscillations in fly-wheels mounted on a shaft, see Reference [9]. Somewhat similar considerations apply to fractionating columns in chemical engineering.

4.7 Vibrations of drum heads

The Bessel function introduced in Section 3.3 also provides the solution to the problem of the vibrations of a drum-head. Section 4.5 was concerned with the vibrations of a cable net. The continuous analogue of this is a vibrating membrane, and it may be shown that the displacements of such a membrane satisfy the differential equation

$$\frac{\partial^2 z}{\partial x^2} + \frac{\partial^2 z}{\partial y^2} = f(x, y), \qquad (4.22)$$

where $f(x, y)$ is the perturbing force at (x, y), which is the continuous analogue of Eq. (4.21).

The normal vibrations of a rectangular membrane are easily calculated, but perhaps more interesting are the vibrations of a circular membrane, or drum-head.

It is shown in calculus books that if the expression

$$\frac{\partial^2 z}{\partial x^2} + \frac{\partial^2 z}{\partial y^2}$$

is put into polar co-ordinates then it takes the form

$$\frac{\partial^2 z}{\partial r^2} + \frac{1}{r}\frac{\partial z}{\partial r} + \frac{1}{r^2}\frac{\partial^2 z}{\partial \theta^2}.$$

We will restrict attention to the simplest type of oscillation. When the displacements at all points equidistant from the centre of the membrane are equal, z depends on r only. The more complicated oscillations in which z is also a function of θ can be handled almost as easily, but we will deliberately deal only with the simplest case. The term involving θ is then zero.

The equation giving the frequencies of the normal modes is easily found to be

$$z'' + \frac{1}{r}z' + \omega^2 z = 0, \qquad (4.23)$$

where primes denote differentiation with respect to r.

There is the obvious boundary condition that z must vanish at the edge of the drum, when $r = 1$ (say), but it is not so obvious what the boundary condition of the differential equation is when $r = 0$. This will emerge later.

We attempt to solve (4.23) by means of the infinite series

$$z = \sum_{n=0}^{\infty} a_n r^n.$$

Differentiating
$$z' = \sum_{n=1}^{\infty} n a_n r^{n-1},$$

and
$$z'' = \sum_{n=2}^{\infty} n(n-1) a_n r^{n-2}.$$

Substituting in to Eq. (4.23)

$$\sum_{n=2}^{\infty} n(n-1)a_n r^{n-2} + \sum_{n=1}^{\infty} na_n r^{n-2} + \omega^2 \sum_{n=0}^{\infty} a_n r^n = 0.$$

To satisfy the equation the left-hand side must vanish identically. The lowest power of r present is r^{-1}, of which the coefficient is a_1. Hence, $a_1 = 0$. The coefficients of all the other powers of r can be covered by a general formula, the coefficient of r^{n-2} being

$$n(n-1)a_n + na_n + \omega^2 a_{n-2}, \qquad n \geqslant 2.$$

For this to be zero

$$n^2 a_n + \omega^2 a_{n-2} = 0.$$

Hence

$$a_n = -\frac{\omega^2}{n^2} a_{n-2}.$$

This means that

$$a_2 = -\frac{\omega^2}{2^2} a_0, \qquad a_4 = -\frac{\omega^2}{4^2} a_2 = \frac{\omega^4}{2^2 \cdot 4^2} a_0, \quad \text{etc.}$$

This gives as a solution any arbitrary multiple, a_0, of the series

$$1 - \left(\frac{\omega r}{2}\right)^2 + \frac{1}{(2!)^2}\left(\frac{\omega r}{2}\right)^4 - \frac{1}{(3!)^2}\left(\frac{\omega r}{2}\right)^6 + \ldots;$$

which, using the definition in Section 3.3, is the Bessel function denoted by $J_0(\omega r)$.

It may be shown that the differential equation has another independent solution which tends to infinity as r tends to zero; but this is clearly inapplicable in this problem and this accounts for the boundary condition at $r = 0$. At this point the solution must stay finite. Observe incidentally that the derivative of the solution is zero at $r = 0$, which is what might be expected intuitively from the behaviour of drum-heads.

The boundary condition at the edge of the drum, $r = 1$, is only satisfied if $J_0(\omega) = 0$, and this introduces the eigenvalues. Tables show that the first few values of ω satisfying this relation are approximately

$$2 \cdot 405, \quad 5 \cdot 520, \quad 8 \cdot 654, \quad 11 \cdot 79, \quad 14 \cdot 93.$$

(These are the numbers which occurred in Example 3.6.)

This means that the frequencies of the fundamental and the overtones of a drum are in the ratios

$$1 : 2 \cdot 3 : 3 \cdot 6 : 4 \cdot 9 : 6 \cdot 2 \ldots,$$

and this explains why a drum does not give a musical note. The overtones are not integral multiples of the fundamental.

In our general terminology the frequencies of the normal modes of the drum have been related to the eigenvalues of the operator

$$\mathscr{D} = -\frac{d^2}{dr^2} - \frac{1}{r}\frac{d}{dr}.$$

The eigenvalues are given by $\lambda = \omega^2$, i.e. they are the squares of the numbers listed above. The eigenfunctions are the Bessel functions $J_0(\omega r)$ for those special values of ω, and these give a snapshot of the displacement of the drum-head along a radius when it is vibrating in the appropriate normal mode.

The list of references includes suggestions for further reading. Dickinson [4] gives many school-level examples on operators, but does not employ them on boundary value problems. Rayleigh's great classic [10] is still very readable, and like some other older books, such as Webster [13], gives significant details which many later books omit. The techniques which have been described in the last two chapters are extensively employed in applied mathematics. The (British) *Journal of the Institute of Mathematics and its Applications* contains many papers describing recent developments, a selection being References [2, 3, 7, 11].

References

1. BRAND, T. and SHERLOCK, A. J., *Matrices, Pure and Applied*, Arnold (1970).
2. BUCHHOLDT, H. A., DAVIES, M. and HUSSEY, M. J. L., 'The Analysis of Cable Nets', *J. Inst. Math. and its Appl.*, **4**, 339–58 (1968).
3. DEAN, P., 'Atomic Vibrations in Solids', *J. Inst. Math. and its Appl.*, **3**, 98–165 (1967).
4. DICKINSON, D. R., *Operators*, Macmillan (1967).
5. FLETCHER, T. J. (ed.) *Some Lessons in Mathematics*, CUP (1964).
6. FORT, T., *Finite Differences*, OUP (1948).
7. GUPTA, K. K., 'Free Vibrations of single branch structural systems', *J. Inst. Math. and its Appl.*, **5**, 351–62 (1969).
8. HODGSON, T., *Applied Mathematics for Engineers*, Vol. 3, Chapman and Hall (1931).
9. PRENTIS, J. M. and LECKIE, F. A., *Mechanical Vibrations*, Longmans (1963).
10. RAYLEIGH, J. W. S., *The Theory of Sound*, Macmillan (1877). Reprinted by Dover (1945).

11. SWEET, R. A., 'Properties of a semi-discrete approximation to the beam equation', *J. Inst. Math. and its Appl.*, **5**, 329–39 (1969).
12. TROPPER, A. M., *Matrix Theory for Electrical Engineering Students*, Harrap (1962).
13. WEBSTER, A. G., *Partial Differential Equations*, Reprinted by Dover (1955).

5
Fourier Series and Symmetric Operators

The previous two chapters have raised many matters which demand rigorous discussion, especially some of the questions concerning the eigenvalues and eigenfunctions of differential equations. As a matter of history, the practical problems of mechanical oscillations and of heat conduction led to a number of deep and difficult branches of pure mathematics. In particular the study of topological vector spaces goes back to these practical problems. This branch of mathematics is regarded by the Bourbaki [2], a school of French mathematicians who write under a collective pseudonym, as the keystone of an arch which rests on the twin pillars of algebra and topology. The Bourbaki are sometimes regarded as the extreme exponents of an excessively pure form of mathematics, but it is interesting to note that the area to which they attach such fundamental importance is an area which grew from the need to discuss certain very practical problems in a mathematically rigorous way.

This chapter is in some ways the most theoretical in this book. However, we make no claims to logical rigour; we attempt merely to indicate in outline how the problems may be approached, and we show in a general way what a rigorous theory has to provide.

5.1 Projection matrices

In Section 2.6 it was shown that the projection of the vector $\mathbf{y}$ in the direction of the unit vector $\mathbf{x}$ is $\mathbf{x}(\mathbf{x}'\mathbf{y})$. Since

$$\mathbf{x}(\mathbf{x}'\mathbf{y}) = (\mathbf{x}\mathbf{x}')\mathbf{y}$$

we may say that $\mathbf{y}$ is projected on to $\mathbf{x}$ by pre-multiplying it by $\mathbf{xx}'$ Matrices of this form we will call *projection matrices*.

Exercise 5.1 (harder)
 If $\mathbf{x}$ is not of unit length it has to be normalized, and the projection then is
$$\mathbf{x}(\mathbf{x}'\mathbf{y})/\mathbf{x}'\mathbf{x} = \mathbf{x}(\mathbf{x}'\mathbf{x})^{-1}\mathbf{x}'\mathbf{y}.$$

More generally, show that if $\mathscr{W}$ is a subspace of a vector space $\mathscr{V}$, and $\mathscr{W}$ has as a basis the linearly independent column vectors $\mathbf{x}_1, \ldots, \mathbf{x}_m$, and

$$\mathbf{X} = \begin{bmatrix} \mathbf{x}_1 & \vdots & \ldots & \vdots & \mathbf{x}_m \end{bmatrix},$$

then the matrix giving the orthogonal projection of $\mathscr{V}$ on $\mathscr{W}$ is $\mathbf{X}(\mathbf{X}'\mathbf{X})^{-1}\mathbf{X}'$.

Construct some numerical examples to verify this in two and three dimensions.

We will now continue the discussion of Example 3.3 of Section 3.1. There it was shown that the matrix

$$\mathbf{A} = \begin{bmatrix} 2 & -1 & 0 \\ -1 & 2 & -1 \\ 0 & -1 & 2 \end{bmatrix}$$

has eigenvalues and eigenvectors

$$\begin{aligned} \lambda_1 &= \omega_1^2 = 2 - \sqrt{2}, & (1, \sqrt{2}, 1)'; \\ \lambda_2 &= \omega_2^2 = 2, & (1, 0, -1)'; \\ \lambda_3 &= \omega_3^2 = 2 + \sqrt{2}, & (1, -\sqrt{2}, 1)'. \end{aligned}$$

These eigenvectors may be normalized by dividing by 2, $\sqrt{2}$ and 2 respectively; and they are readily seen to be mutually orthogonal, but this holds in any case by the general Theorem 2.3.

If a mechanical system vibrates, starting from rest with certain arbitrary displacements, then it may be useful to know to what extent the different modes are present in the vibration. This can be found by projecting the vector denoting the initial state on to the normalized eigenvectors. The three projection matrices are, by the result at the head of this section,

$$\begin{bmatrix} \tfrac{1}{2} \\ \tfrac{1}{2}\sqrt{2} \\ \tfrac{1}{2} \end{bmatrix} \begin{bmatrix} \tfrac{1}{2} & \tfrac{1}{2}\sqrt{2} & \tfrac{1}{2} \end{bmatrix} = \begin{bmatrix} \tfrac{1}{4} & \tfrac{1}{4}\sqrt{2} & \tfrac{1}{4} \\ \tfrac{1}{4}\sqrt{2} & \tfrac{1}{2} & \tfrac{1}{4}\sqrt{2} \\ \tfrac{1}{4} & \tfrac{1}{4}\sqrt{2} & \tfrac{1}{4} \end{bmatrix} = \mathbf{A}_1 \quad \text{(say),}$$

$$\begin{bmatrix} \tfrac{1}{2}\sqrt{2} \\ 0 \\ -\tfrac{1}{2}\sqrt{2} \end{bmatrix} \begin{bmatrix} \tfrac{1}{2}\sqrt{2} & 0 & -\tfrac{1}{2}\sqrt{2} \end{bmatrix} = \begin{bmatrix} \tfrac{1}{2} & 0 & -\tfrac{1}{2} \\ 0 & 0 & 0 \\ -\tfrac{1}{2} & 0 & \tfrac{1}{2} \end{bmatrix} = \mathbf{A}_2,$$

$$\begin{bmatrix} \tfrac{1}{2} \\ -\tfrac{1}{2}\sqrt{2} \\ \tfrac{1}{2} \end{bmatrix} \begin{bmatrix} \tfrac{1}{2} & -\tfrac{1}{2}\sqrt{2} & \tfrac{1}{2} \end{bmatrix} = \begin{bmatrix} \tfrac{1}{4} & -\tfrac{1}{4}\sqrt{2} & \tfrac{1}{4} \\ -\tfrac{1}{4}\sqrt{2} & \tfrac{1}{2} & -\tfrac{1}{4}\sqrt{2} \\ \tfrac{1}{4} & -\tfrac{1}{4}\sqrt{2} & \tfrac{1}{4} \end{bmatrix} = \mathbf{A}_3,$$

respectively.

Then any arbitrary vector $\mathbf{y}$ is resolved into its components along the directions of the eigenvectors of $\mathbf{A}$ by the expressions

$$\mathbf{y} = \mathbf{A}_1 \mathbf{y} + \mathbf{A}_2 \mathbf{y} + \mathbf{A}_3 \mathbf{y} = \mathbf{x}_1(\mathbf{x}_1'\mathbf{y}) + \mathbf{x}_2(\mathbf{x}_2'\mathbf{y}) + \mathbf{x}_3(\mathbf{x}_3'\mathbf{y}).$$

The latter relation expresses **y** as the sum of suitable multiples of the eigenvectors of **A**, and this is an idea that will occupy us for most of the rest of the chapter.

Exercise 5.2

Projection matrices have a number of interesting properties. Verify in this case that

$$\mathbf{A}_r \mathbf{A}_s = 0 \quad \text{if } r \neq s,$$
$$= \mathbf{A}_r \quad \text{if } r = s;$$

that

$$\mathbf{A}_1 + \mathbf{A}_2 + \mathbf{A}_3 = \mathbf{I};$$

that

$$\lambda_1 \mathbf{A}_1 + \lambda_2 \mathbf{A}_2 + \lambda_3 \mathbf{A}_3 = \mathbf{A};$$

and hence that

$$\lambda_1^r \mathbf{A}_1 + \lambda_2^r \mathbf{A}_2 + \lambda_3^r \mathbf{A}_3 = \mathbf{A}^r.$$

In particular verify that

$$\frac{1}{\lambda_1} \mathbf{A}_1 + \frac{1}{\lambda_2} \mathbf{A}_2 + \frac{1}{\lambda_3} \mathbf{A}_3 = \mathbf{A}^{-1} = \mathbf{G}.$$

(**G** is the flexibility matrix.)

Show that these results hold for any symmetric matrix with distinct eigenvalues. (*Hint*: Theorem 2.4 is useful.) The results may be adapted for symmetric matrices with repeated eigenvalues, but this is harder to prove. The corresponding results for matrices which are not necessarily symmetric are given in Section 6.6.

Exercise 5.3

Show that the projection matrices may also be put in the form

$$\mathbf{A}_1 = (\mathbf{A} - \lambda_2 \mathbf{I})(\mathbf{A} - \lambda_3 \mathbf{I})/(\lambda_1 - \lambda_2)(\lambda_1 - \lambda_3), \text{ etc.}$$

(Compare the Lagrange interpolation polynomials of Section 1.3.)

Exercise 5.4

Show that if any matrix **A** has an eigenvalue λ then $\mathbf{A}^2$ has an eigenvalue λ^2 and $\mathbf{A}^r$ has an eigenvalue λ^r. Show further that if $f(\mathbf{A})$ denotes any polynomial function of the matrix **A**, that $f(\mathbf{A})$ has an eigenvalue $f(\lambda)$, and that $f(\mathbf{A}) = 0$ implies that $f(\lambda) = 0$.

Projection matrices have the property that $\mathbf{A}^2 = \mathbf{A}$. Deduce that all of their eigenvalues are either 0 or 1.

5.2 Fourier series

The above results for matrix operators now have to be extended to differential operators. We will discuss first of all the case of the vibrating string in Example 3.5. The attempt to express an arbitrary displacement of the string as the sum of eigenfunctions gives rise to Fourier series—an instrument of great importance in applied mathematics. This section is concerned only with certain special Fourier series, which use the eigenfunctions of the operator $-d^2/dx^2$, with the boundary conditions $\eta(0) = \eta(1) = 0$; but the subsequent section is concerned with series which use the more general eigenfunctions which arise with other operators, as in Examples 3.6 and 4.1.

In elementary textbooks Fourier sine series are usually explained much as follows. We require the basic integrals

$$\int_0^1 \sin n\pi x \sin m\pi x \, dx$$

$$= \tfrac{1}{2} \int_0^1 \cos(m-n)\pi x \, dx - \tfrac{1}{2} \int_0^1 \cos(m+n)\pi x \, dx,$$

$$= \frac{1}{2(m-n)\pi} \Big[\sin(m-n)\pi x\Big]_0^1 - \frac{1}{2(m+n)\pi}\Big[\sin(m+n)\pi x\Big]_0^1,$$

$$= 0; \qquad (5.1)$$

and

$$\int_0^1 \sin^2 n\pi x \, dx$$

$$= \tfrac{1}{2}\int_0^1 (1 - \cos 2n\pi x) \, dx,$$

$$= \tfrac{1}{2}\Big[x - \frac{1}{2n\pi}\sin 2n\pi x\Big]_0^1,$$

$$= \tfrac{1}{2}. \qquad (5.2)$$

If we assume that some function $\eta(x)$ can be expressed in the form

$$\eta(x) = a_1 \sin \pi x + a_2 \sin 2\pi x + \ldots, \qquad 0 < x < 1,$$

then making certain further assumptions, multiplying by $\sin n\pi x$ and integrating from $x = 0$ to $x = 1$, we have

$$\int_0^1 \eta(x) \sin n\pi x \, dx = \tfrac{1}{2} a_n, \qquad (5.3)$$

since all of the other terms on the right-hand side vanish.

Hence

$$a_n = 2\int_0^1 \eta(x) \sin n\pi x \, dx, \qquad n > 0.$$

We thus expect that

$$\eta(x) = \sum_{n=1}^{\infty} a_n \sin n\pi x, \qquad 0 < x < 1.$$

This series is called the Fourier sine series of $\eta(x)$. There are also Fourier cosine series, and the cosines may be regarded as the eigenfunctions arising from a similar problem with different boundary conditions.

The above derivation of the *Fourier coefficients*, the terms $a_1, a_2, \ldots$, makes assumptions about the convergence of series and the validity of such operations as the term by term integration of a series which are not always justified. It is possible to state fairly simple restrictions on the function f which are sufficient to ensure the validity of the processes, but such simple conditions are by no means always satisfied in problems of practical interest. Indeed the search for mathematical theories adequate for commonly arising problems has led to the development of extensive areas of difficult pure mathematics. Fourier himself was not able to give a satisfactory mathematical justification for his methods, and to do so would be beyond the scope of this book. Our aim is to indicate something of the rich variety of problems which can be solved if an adequate theory can be constructed.

In a sense it is possible to regard $(a_1, a_2, a_3, \ldots)$ as the components of the vector $\eta(x)$, with respect to a co-ordinate system in which the functions

$$\sin \pi x, \ \sin 2\pi x, \ \sin 3\pi x, \ \ldots$$

are the base vectors.

It is helpful to normalize, that is to arrange that the base vectors have unit length, by taking instead

$$g_1(x) = \sqrt{2} \sin \pi x, \qquad g_2(x) = \sqrt{2} \sin 2\pi x, \text{ etc.}$$

Then these functions satisfy the relations

$$\int_0^1 g_r(x) g_s(x) \, dx = 0, \qquad r \neq s,$$
$$= 1, \qquad r = s, \tag{5.4}$$

which correspond to the inner product relations between a set of orthogonal unit vectors in a vector space; but instead of forming inner products as we did with vectors we now form integrals.

Then if $\eta(x)$ can be expressed in the form

$$\eta(x) = \sum_{n=1}^{\infty} a_n g_n(x),$$

we have
$$a_n = \int_0^1 \eta(x) g_n(x) \, dx,$$
and this corresponds to forming the inner product of some arbitrary vector with a base vector to find its components. It projects the function in a particular direction. Certain familiar results extend into the new circumstances. Thus (ignoring problems of convergence)
$$\int_0^1 \eta^2(x) \, dx = a_1^2 + a_2^2 + a_3^2 + \cdots$$
can be regarded as an extension of Pythagoras's theorem.

The notation we have been employing is traditional, but it is a little unfortunate because it obscures certain resemblances to which we wish to draw attention. The vector **x** has components $x_1, \ldots, x_n$; i.e it has a component corresponding to each integer r, $1 \leqslant r \leqslant n$. On the other hand the function η has a value for each value of x in the interval $(0, 1)$; i.e. it has a value corresponding to each x, $0 < x < 1$. The inconvenience in this traditional notation is that in the first case the 'x' is suggesting the *dependent variable* and in the second case it suggests the *independent variable*. The beginner is expected to take this in his stride!

The applications of the Fourier sine series to the vibrating string problem are described in many mathematics and physics texts. In particular, expressions of the form
$$\eta(x, t) = \sum_{n=1}^{\infty} a_n g_n(x) \sin n\pi t$$
are useful for solving problems in which the string is in its undisplaced position at $t = 0$, and is set in motion with a prescribed distribution of velocities. For problems in which the string is released from rest in a displaced position Fourier cosine series are required.

Special considerations arise when, for some reason, a finite number of terms of the Fourier sine series is all that is required. Instead of using the integral formulae (5.1) and (5.2), finite sums may be used instead. We have the trigonometric identities

$$\sum_{t=1}^{N} \sin \frac{tn\pi}{N+1} \sin \frac{tm\pi}{N+1} = 0, \qquad n \neq m, \qquad (5.5)$$

$$= \tfrac{1}{2} N, \qquad n = m. \qquad (5.6)$$

These equations may be proved as exercises in trigonometric manipulation, but they may also be seen as expressing the orthogonality of the eigenfunctions in Example 4.3, Eq. (4.14). Apart from a numerical constant $N + 1$ arising from the units used, Eqs. (5.5) and (5.6) may be regarded as approximations to Eqs. (5.1) and (5.2), evaluating the integrals by taking sample values at the points

$$\frac{1}{N+1}, \frac{2}{N+1}, \ldots, \frac{N}{N+1}.$$

If now

$$\eta(x) = \sum_{n=1}^{N} a_n \sin n\pi x,$$

listing the separate equations obtained by putting $x = t/(N+1)$ for $t = 1, \ldots, N$ we have

$$\eta\left(\frac{1}{N+1}\right) = a_1 \sin \frac{\pi}{N+1} + \ldots + a_N \sin \frac{N\pi}{N+1},$$

$$\eta\left(\frac{2}{N+1}\right) = a_1 \sin \frac{2\pi}{N+1} + \ldots + a_N \sin \frac{2N\pi}{N+1}, \text{ etc., down to}$$

$$\eta\left(\frac{N}{N+1}\right) = a_1 \sin \frac{N\pi}{N+1} + \ldots + a_N \sin \frac{N^2\pi}{N+1}.$$

Multiplying these in turn by

$$\sin \frac{m\pi}{N+1}, \sin \frac{2m\pi}{N+1}, \ldots, \sin \frac{Nm\pi}{N+1}$$

and adding, most of the terms on the right-hand side go out, the sums of columns vanishing by Eq. (5.5). We are left with

$$\sum_{t=1}^{N} \eta\left(\frac{t}{N+1}\right) \sin \frac{tm\pi}{N+1} = \tfrac{1}{2} N a_m.$$

This relation corresponds to Eq. (5.3), and it enables the required coefficient a_m to be found.

The interesting feature of the calculation is that the integrals are replaced by sums, the terms involving the values of the functions at certain points, these points being the zeros of the next eigenfunction, which in this case is $\sin (N+1)\pi x$. Similar relations may be expected to hold for the more general systems of eigenfunctions to be studied in the next section, but the commonly available books give very little information on this point.

A further important property of finite Fourier series is discussed in Section 5.4.

5.3 Chebyshev polynomials

Experience shows that Fourier series can provide very useful approximations to functions over a given range. For some applications sine series are most suitable, for others cosines are used. Cosine series can be handled in a very similar way to the sine series used above. There are various well-known trigonometrical identities which enable functions such as $\cos n\theta$ and $\sin n\theta$ to be expressed in terms of $\cos \theta$ and $\sin \theta$.

Thus we have
$$\cos 2\theta = 2\cos^2 \theta - 1$$
and
$$\cos 3\theta = 4\cos^3 \theta - 3\cos \theta.$$

Since
$$\cos(n+1)\theta + \cos(n-1)\theta = 2\cos\theta \cos n\theta \qquad (5.7)$$
it follows that, putting first $n = 3$ and then $n = 4$,
$$\cos 4\theta = 8\cos^4 \theta - 8\cos^2 \theta + 1,$$
and
$$\cos 5\theta = 16\cos^5 \theta - 20\cos^3 \theta + 5\cos \theta.$$

In fact, by working a step at a time, $\cos n\theta$ may be expressed as a polynomial of degree n in $\cos \theta$, for any n. Putting $x = \cos \theta$, call this polynomial $T_n(x)$, so that
$$T_n(x) = \cos(n \cos^{-1} x), \qquad -1 \leqslant x \leqslant 1.$$

Rewriting the above equations we have
$$T_0(x) = 1, \qquad T_1(x) = x, \qquad T_2(x) = 2x^2 - 1,$$
$$T_3(x) = 4x^3 - 3x, \qquad T_4(x) = 8x^4 - 8x^2 + 1,$$
$$T_5(x) = 16x^5 - 20x^3 + 5x.$$

These polynomials have applications in numerical approximation which will now be explained. Since $\cos n\theta$ has maxima and minima with values ± 1 at $\theta_r = r\pi/n$, $r = 0, 1, \ldots, n$, $T_n(x)$ is a polynomial of degree n with maxima and minima at
$$x_r = \cos \frac{r\pi}{n}, \qquad r = 0, 1, \ldots, n \; ;$$
and over the interval $-1 \leqslant x \leqslant 1$ its absolute value does not exceed unity.

It is quite easy to show by induction, using Eq. (5.7), that the coefficient of x^n in $T_n(x)$ is 2^{n-1}, when $n \geqslant 1$.

Hence
$$p_{n-1}(x) = x^n - \frac{1}{2^{n-1}} T_n(x) \qquad (5.8)$$

is a polynomial of degree $n-2$ which differs from x^n by a known amount; in fact over the interval $-1 \leq x \leq 1$ it differs from x^n by an amount whose magnitude does not exceed $1/2^{n-1}$. It can be shown furthermore (Reference [8]) that there is no polynomial of degree less than n which provides a better approximation than this, but we will not do so here. Equation (5.8) is therefore useful when seeking polynomial approximations of lower degree. For example

$$e^x = 1 + x + x^2/2! + x^3/3! + x^4/4! + \cdots$$
$$p_3(x) = x^4 - \tfrac{1}{8}T_4(x) = x^2 - \tfrac{1}{8}.$$

The difference between x^4 and $x^2 - \tfrac{1}{8}$ does not exceed $\tfrac{1}{8}$ over the range $-1 \leq x \leq 1$. Therefore, over this range if we are prepared to truncate the series for e^x at the term in x^4, we may replace $x^4/4!$ by $(x^2 - \tfrac{1}{8})/4!$ and by so doing introduce a further error no greater than $1/8.4!$ $= 0.00521$. This gives the approximation

$$e^x = \frac{191}{192} + x + \frac{13}{24}x^2 + \frac{1}{6}x^3 \text{ (approx.)}$$

This is a better approximation to e^x over the range $-1 \leq x \leq 1$ than the expression obtained by truncating the power series after the term in x^3. The removal of the term in x^4 has been compensated for, as much as possible, by adjusting the term in x^2 and the constant.

Exercise 5.5
Make up some similar examples for yourself.

Chebyshev polynomials have further applications in numerical analysis which arise from the suitability of these polynomials for numerical interpolation and from the properties of rapid convergence which series of them usually have [9, 11, 12]. This illustrates further the points made in Section 1.3, where it was seen that differing sets of polynomials have advantages as a basis for the vector space of polynomials, according to the practical problems being tackled.

5.4 The Sturm–Liouville equation and symmetric operators

Examples 3.5, 3.6 and 4.1, problems concerned with the vibration of one-dimensional continuous systems under various conditions, all led to equations of the form

$$\ddot{\eta} = (T(x)\eta')',$$

where $\eta(x)$ is the displacement of the string (or chain) at the point x, and $T(x)$ is the tension in the string at the point x. (Restricting attention to cases where there are no externally applied forces.) More general problems may arise in which the density of the string varies along its length, being $\rho(x)$ at the point x. In these cases $\ddot{\eta}$ has to be replaced by $\rho(x)\ddot{\eta}$. The next step in the search for the normal modes in all of these problems is to put $\eta = y(x) \sin kt$, and after cancelling the factor $\sin kt$ we arrive at the equation

$$(T(x)y')' + k^2\rho(x)y = 0. \tag{5.9}$$

There are also boundary conditions on y at two points which depend on the particular problem. In the cases when the string or chain is fastened to a rigid peg the boundary condition is $y = 0$. When, as in Example 3.7, the fastening is to a peg which can be elastically displaced the boundary condition is of the more general form $\alpha y + \beta y' = 0$. In some of our previous problems the end of the chain was free; then calculations were assisted by the fact that $T(x)$ was zero at the free end, and this can take the place of the other forms of boundary condition. The angular frequencies of the normal modes of vibration are given by the eigenvalues (for k) of Eq. (5.9), and the patterns of the various normal modes are given by the corresponding eigenfunctions.

Similar considerations arise in problems on the conduction of heat in metal bars (one of the simplest cases to consider theoretically). These problems were investigated by Fourier, and Fourier series were developed with this application as the stimulus. Eigenvalues and eigenfunctions play a big part in the theory of heat conduction, but we have not selected examples from this field because there is no immediate, intuitively clear physical interpretation of eigenfunctions as there is with mechanical vibrations. Those with some knowledge of the physics of heat conduction might agree that the following equations offer the foundations of a plausible theory. Others may be interested to see them, if only as an historical exhibit.

The conduction of heat in an inhomogeneous bar is described by the equation

$$[k(x)y']' + [\lambda g(x) - l(x)]y = 0, \tag{5.10}$$

with boundary conditions at $x = a$ and $x = b$ of the form

$$y'(a) - hy(a) = 0, \quad y'(b) + Hy(b) = 0;$$

where $y(x)$ is the temperature at the point x,
$\quad k(x)$ is the thermal conductivity at the point x,
$\quad g(x)$ is the specific heat of the bar at the point x,
$\quad l(x)$ is the emissivity at the surface at the point x,
and h and H are the emissivities at the ends of the bar.

The derivation of this equation depends on no more than the knowledge that the flow of heat in the bar is proportional to the temperature gradient, the laws of emission from a hot surface, and the overall conservation of heat.

Equations (5.9) and (5.10) are both covered by the equation

$$[p(x)y']' + [q(x) + \lambda r(x)]y = 0, \quad a \leqslant x \leqslant b, \quad (5.11)$$

with boundary conditions of the form $\alpha y + \beta y' = 0$ at a and b, or with $p(x)$ vanishing at a and/or b.

This equation was studied by Sturm and Liouville (separately) in 1836 and it is called the Sturm–Liouville equation. Something of the history of the equation may be read in Coppel [6] and in Bourbaki [2].

Books on differential equations, such as those by Burkill [4] and Ince [10], derive a number of properties of the eigenfunctions of this equation; but these elegant properties are for the most part no more than the mathematical formulation of physical properties which are intuitively plain and straightforward once the behaviour of the vibrating mechanical systems has been appreciated. Thus we may quote, without proving, the property that the Sturm–Liouville equation has solutions which fit the boundary conditions only for a sequence of discrete values (the eigenvalues) of the parameter λ. With suitable numbering the nth eigenfunction has $n-1$ zeros in the interval $a < x < b$, and the zeros of the nth eigenfunction separate the zeros of the $(n+1)$th. There are analogous interlacing properties for the eigenvectors of the matrices which describe the discrete vibrating systems, but it may be observed that textbooks say very little about these.

In many of the applications $r(x)$ of Eq. (5.11) is unity throughout, and in any case it is possible to use a change of variable to transform the equation to a new equation of the same form but with $r(x) = 1$. So to simplify the presentation we will discuss the theory of the Sturm–Liouville equation with $r(x) = 1$; but the more general theory is not really any more difficult.

One important property of the eigenfunctions of the Sturm–Liouville equation is that they are orthogonal, as was seen previously in the particular case of the vibrating uniform string.

Let λ_1 and λ_2 be two eigenvalues of the equation and $y_1(x)$ and $y_2(x)$ the corresponding eigenfunctions. Then, remembering that we are taking $g(x) = 1$,

$$(py_1')' + (q + \lambda_1)y_1 = 0,$$

and

$$(py_2')' + (q + \lambda_2)y_2 = 0.$$

Multiplying these equations by $y_2(x)$ and $y_1(x)$ respectively, and subtracting

$$y_2(py_1')' - y_1(py_2')' = (\lambda_2 - \lambda_1)y_1 y_2.$$

Integrating by parts from a to b we have

$$\left[p(y_1'y_2 - y_1y_2')\right]_a^b = (\lambda_2 - \lambda_1)\int_a^b y_1y_2\,dx,$$

the remaining terms on the left-hand side cancelling out. Whichever forms of boundary conditions apply, the expression on the left vanishes at $x = a$ and also at $x = b$. (Check this for the various cases.) If $\lambda_1 \neq \lambda_2$ we therefore have

$$\int_a^b y_1y_2\,dx = 0, \tag{5.12}$$

and the eigenfunctions are orthogonal, in the sense that this integral of their product vanishes.

This orthogonality of the eigenfunctions is to be seen as an analogue for differential equations of Theorem 2.3 in Section 2.8, which demonstrates the orthogonality of eigenvectors of a symmetric matrix. The differential operator

$$\frac{d}{dx}\left[p(x)\frac{d}{dx}\right]$$

in the Sturm–Liouville equation is called *self-adjoint* or *symmetric*, but it is a little too early as yet to see what these words mean.

It will be convenient to state the result as a theorem.

THEOREM 5.1
Any two eigenfunctions associated with distinct eigenvalues of the Sturm–Liouville equation are orthogonal, in the sense that the integral of their product over the interval (a, b) is zero.

We will now show how the results in both systems, discrete systems with matrices and continuous systems with differential operators, can be obtained within the framework of a single theory. This will also show how mathematics often advances. Correspondences are noticed between different theories; but analogies are never enough to establish a mathematical argument—although they may suggest one. Something stronger than analogy has to be established. It is necessary to set up a more abstract mathematical system which includes the two previous theories as special cases.

For this purpose we need the ideas of Chapter 1 on linear spaces composed of *points* or *vectors* on which we may carry out two operations—addition and scalar multiplication. When we are concerned with differential operators the function spaces involved are of the kind described in Section 3.3. We also need to introduce into our space an *inner product*, as was done in a purely geometrical context in Chapter 2.

THE STURM–LIOUVILLE EQUATION

What is important in subsequent proofs is the algebraic properties which the inner product possesses. The inner product will mean one thing in the vector spaces of finite dimension with which we are familiar and it will mean another thing in the function spaces we have begun to explore. But the algebraic properties of the inner product will be the same in the two cases, and it is these that matter.

To assist the discussion we will adopt a new notation for the inner product, writing the inner product of vectors **u** and **v** as [**u**, **v**],

$$[\mathbf{u}, \mathbf{v}] = \sum_r u_r v_r = \mathbf{u}' \mathbf{v}.$$

In function spaces

$$[\mathbf{u}, \mathbf{v}] = \int_a^b u(x)\, v(x)\, dx.$$

Note that it is very easy to write the Rayleigh quotient of an operator $\mathscr{D}$ in inner product notation. It is $[\mathbf{y}, \mathscr{D}\mathbf{y}]/[\mathbf{y}, \mathbf{y}]$.

Exercise 5.6

Prove that in each of these types of space the inner product has the following properties.

(i) $[\mathbf{u}, \mathbf{v}] = [\mathbf{v}, \mathbf{u}]$,
(ii) $[\mathbf{u}, k\mathbf{v}] = k[\mathbf{u}, \mathbf{v}]$, where k is any scalar multiplier,
(iii) $[\mathbf{u}, \mathbf{v} + \mathbf{w}] = [\mathbf{u}, \mathbf{v}] + [\mathbf{u}, \mathbf{w}]$,
(iv) $[\mathbf{u} + \mathbf{v}, \mathbf{u} + \mathbf{v}] = [\mathbf{u}, \mathbf{u}] + 2[\mathbf{u}, \mathbf{v}] + [\mathbf{v}, \mathbf{v}]$.

If we now try to write out the two proofs of the orthogonality of eigenvectors (Theorem 2.3 in Section 2.8) and eigenfunctions (Theorem 5.1 above) in a common notation we may proceed as follows. Suppose that

$$\mathscr{A}\mathbf{y}_1 = \lambda_1 \mathbf{y}_1,$$
$$\mathscr{A}\mathbf{y}_2 = \lambda_2 \mathbf{y}_2,$$

where we write $\mathscr{A}$ as an operator in our general space. In the matrix case $\mathscr{A}$ means multiply by the matrix **A**, in the case of the differential equations $\mathscr{A}$ corresponds to the differential operators which have previously been denoted by $\mathscr{D}$.

Attempting to continue the proof, using the above equations and forming inner products

$$[\mathbf{y}_2, \mathscr{A}\mathbf{y}_1] = \lambda_1[\mathbf{y}_2, \mathbf{y}_1],$$

and

$$[\mathbf{y}_1, \mathscr{A}\mathbf{y}_2] = \lambda_2[\mathbf{y}_1, \mathbf{y}_2].$$

Now for the proof to continue we need, for any two vectors $\mathbf{y}_1$ and $\mathbf{y}_2$,

$$[\mathbf{y}_1, \mathscr{A}\mathbf{y}_2] = [\mathbf{y}_2, \mathscr{A}\mathbf{y}_1]. \tag{5.13}$$

Is this relation satisfied? Whilst $[\mathbf{y}_1, \mathbf{y}_2] = [\mathbf{y}_2, \mathbf{y}_1]$ we have no grounds for assuming that Eq. (5.13) holds for all the operators $\mathscr{A}$ which we are likely to come across.

In the matrix case

$$[\mathbf{y}_2, \mathbf{A}\mathbf{y}_1] = \mathbf{y}_2'\mathbf{A}\mathbf{y}_1,$$
$$= \mathbf{y}_1'\mathbf{A}'\mathbf{y}_2, \text{ (transposing the } 1 \times 1 \text{ matrix)}.$$
$$= [\mathbf{y}_1, \mathbf{A}'\mathbf{y}_2],$$

and if $\mathbf{A}' = \mathbf{A}$ then this is the same as $[\mathbf{y}_1, \mathbf{A}\mathbf{y}_2]$. Hence the orthogonal property of the eigenvectors only holds (in general) for symmetric matrices. If $\mathbf{A}' = \mathbf{A}$ the proof continues just as before:

$$(\lambda_1 - \lambda_2)[\mathbf{y}_1, \mathbf{y}_2] = 0,$$

and so if $\lambda_1 \neq \lambda_2$ then $[\mathbf{y}_1, \mathbf{y}_2] = 0$.

How does this apply in the case of the differential equation? To conform to earlier notation the operator $\mathscr{D}$ must be defined by

$$\mathscr{D}y = -[p(x)y']' - q(x)y.$$

This time we have to prove that

$$[\mathbf{y}_2, \mathscr{D}\mathbf{y}_1] = [\mathbf{y}_1, \mathscr{D}\mathbf{y}_2].$$

(In the 'space' context we continue to print vectors in bold face type.) By definition

$$\left[\mathbf{y}_2, \mathscr{D}\mathbf{y}_1\right] = -\int_a^b y_2\left[\left(py_1'\right)' + qy_1\right]dx$$
$$= -\left[py_1' y_2\right]_a^b + \int_a^b \left(py_1' y_2' - qy_1 y_2\right) dx.$$

Using the boundary conditions the first term disappears and we are left with the second term only, which is symmetric in the suffixes 1, 2. Hence $[\mathbf{y}_1, \mathscr{D}\mathbf{y}_2]$ reduces to the same thing and the required result is established. For any two functions, $y_1(x)$ and $y_2(x)$, in the space

$$[\mathbf{y}_1, \mathscr{D}\mathbf{y}_2] = [\mathbf{y}_2, \mathscr{D}\mathbf{y}_1].$$

This property is described by saying that $\mathscr{D}$ is a *symmetric operator*. This explains the importance of the standard form chosen for the Sturm–Liouville equation. The operator has this property of symmetry, which differential operators in general do not. The two theorems on eigenvectors of symmetric matrices and on eigenfunctions of Sturm–Liouville systems are summarized in the one result.

THEOREM 5.2

Any eigenvectors associated with distinct eigenvalues of a symmetric operator in a linear space are orthogonal.

Symmetric operators involving integration instead of differentiation will appear later. The theorem will then hold automatically for these as

THE STURM–LIOUVILLE EQUATION

well. We will now deduce further properties from the symmetry of the operator. It is interesting that this property alone is strong enough to prove the results which follow. When the properties of Sturm–Liouville systems are discussed in books on the calculus, integration by parts is used frequently. On the approach here it is used once to prove the symmetry of the operator. This effectively algebraizes the problem, and thereafter algebraic manipulation takes the place of integrations.

Some further results may be established.

THEOREM 5.3

If $\mathscr{A}$ is a real symmetric operator and $\mathbf{y}$ is any vector, possibly complex, then $[\bar{\mathbf{y}}, \mathscr{A}\mathbf{y}]$ is real. (The bar denotes the complex conjugate.)

Proof

Put $\quad w = [\bar{\mathbf{y}}, \mathscr{A}\mathbf{y}]$,

then $\quad \bar{w} = [\mathbf{y}, \mathscr{A}\bar{\mathbf{y}}]$, since $\mathscr{A}$ is real,
$\quad\quad\quad = [\bar{\mathbf{y}}, \mathscr{A}\mathbf{y}]$, using the symmetry of $\mathscr{A}$,
$\quad\quad\quad = w$.

Hence w is real.

COROLLARY

It follows from this that all the eigenvalues of $\mathscr{A}$ are real.

Because if $\quad\quad\quad \mathscr{A}\mathbf{y} = \lambda \mathbf{y}$,

then $\quad\quad\quad [\bar{\mathbf{y}}, \mathscr{A}\mathbf{y}] = \lambda [\bar{\mathbf{y}}, \mathbf{y}]$,

and λ is the ratio of two real numbers.

In any linear space we may attempt to expand an arbitrary vector in terms of the eigenvectors of a symmetric operator in the same way. Working just as with Fourier series, if

$$\mathbf{y} = a_1 \mathbf{y}_1 + a_2 \mathbf{y}_2 + a_3 \mathbf{y}_3 + \ldots$$

then $\quad\quad\quad [\mathbf{y}_r, \mathbf{y}] = a_r [\mathbf{y}_r, \mathbf{y}_r]$

since $\mathbf{y}_r$ is orthogonal to all the other eigenvectors. If in addition we normalize each eigenvector, i.e. adjust the scale so that $[\mathbf{y}_r, \mathbf{y}_r] = 1$,

then $\quad\quad\quad a_r = [\mathbf{y}_r, \mathbf{y}]$

and we have the expansion

$$\mathbf{y} = \sum_r [\mathbf{y}_r, \mathbf{y}] \mathbf{y}_r.$$

It is necessary to give a very clear warning. This completely ignores the many difficulties of convergence which arise when these ideas are applied to spaces of infinite dimension. The difficulties encountered in advanced treatments of this topic nearly all centre around this problem of convergence.

The inner products $[\mathbf{y}_r, \mathbf{y}]$ occurring in the above formula may be termed the generalized Fourier coefficients of $\mathbf{y}$.

Exercise 5.7

If the functions $y(x)$ and $z(x)$ have generalized Fourier coefficients $(a_1, a_2, ...)$ and $(b_1, b_2, ...)$ respectively, show that

$$[\mathbf{y}, \mathbf{z}] = a_1 b_1 + a_2 b_2 + \ldots .$$

(Difficulties of convergence may be ignored.)

We have already discussed one meaning which can be attached to the idea that the differential operator appearing in the Sturm–Liouville equation is *symmetric* or *self-adjoint* to employ the term which is most common in the context of differential equations.

In books concerned purely with the theory of differential equations something of the following kind is usually stated. The equation

$$p(x)y'' + q(x)y' + r(x)y = 0 \tag{5.14}$$

is said to be *self-adjoint* if $q(x) = p'(x)$. This definition has the immediate corollary that the equation may be written in the form

$$(p(x)y')' + r(x)y = 0,$$

but we are given no insight at all as to why this should matter.

There are other points of view. We saw earlier in Theorem 3.1 that the symmetry of the flexibility matrix arises from the conservation of energy. The same is true, of course, for the stiffness matrix. It would be possible to show that the conservation of energy imposes self-adjointness on the differential operators which describe energy-conserving vibrating systems, but we will take another course.

In Section 4.3 difference equations arose from the consideration of sets of discrete particles. It is also possible to derive difference equations by starting with the differential equations of continuous media and attempting to approximate to them. In this way we may consider the solution in the neighbourhood of some point x_0 and proceed by steps of magnitude h. It is a notational convenience to write

$$x_n = x_0 + nh \quad \text{and} \quad y_n = y(x_0 + nh).$$

The aim is now to find the values of y_n at the various points x_n (see Fig. 5.1).

THE STURM–LIOUVILLE EQUATION

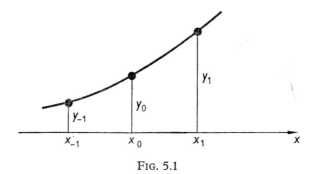

FIG. 5.1

It is plausible to approximate to $y' = dy/dx$ by taking

$$y'_0 = (y_1 - y_{-1})/2h.$$

and similarly at other points.

$(y_1 - y_0)/h$ is the approximate gradient at the point $x_0 + \tfrac{1}{2}h$ and $(y_0 - y_{-1})/h$ is the approximate gradient at the point $x_0 - \tfrac{1}{2}h$. Therefore, over a distance h the gradient increases by an amount

$$(y_1 - 2y_0 + y_{-1})/h,$$

so the rate of change of the gradient at x_0 is given approximately by taking

$$y''_0 = (y_1 - 2y_0 + y_{-1})/h^2,$$

and similarly at other points. As an approximation to Eq. (5.14) we may therefore take at the point x_0

$$p_0(y_1 - 2y_0 + y_{-1}) + \tfrac{1}{2}hq_0(y_1 - y_{-1}) + h^2 r_0 y_0 = 0,$$

or $\quad y_{-1}(p_0 - \tfrac{1}{2}hq_0) - y_0(2p_0 - h^2 r_0) + y_1(p_0 + \tfrac{1}{2}hq_0) = 0.$

Similar equations with different suffixes apply at the other points x_n. This set of difference equations has a banded matrix, that is a matrix in which all of the terms which are not on the principal diagonal or the diagonals immediately above and below are zero. (The numbering of the rows and columns of the matrix will depend on the values of x which it is necessary to cover, and some adjustment may be necessary at the corners of the matrix in order to accommodate the boundary conditions properly.) Our concern is only with a typical situation on the diagonal, which is exemplified by

$p_{-1}-\tfrac{1}{2}hq_{-1}$	$-(2p_{-1}-h^2 r_{-1})$	$p_{-1}+\tfrac{1}{2}hq_{-1}$	.	.
.	$p_0-\tfrac{1}{2}hq_0$	$-(2p_0-h^2 r_0)$	$p_0+\tfrac{1}{2}hq_0$	.
.	.	$p_1-\tfrac{1}{2}hq_1$	$-(2p_1-h^2 r_1)$	$p_1+\tfrac{1}{2}hq_1$

This matrix is symmetric if, and only if

$$p_0 - \tfrac{1}{2}hq_0 = p_{-1} + \tfrac{1}{2}hq_{-1}$$

and
$$p_1 - \tfrac{1}{2}hq_1 = p_0 + \tfrac{1}{2}hq_0,$$

and similarly for all the other suffixes. Adding the equations and rearranging

$$p_1 - p_{-1} = hq_0 + \tfrac{1}{2}h(q_{-1} + q_1).$$

To the order of accuracy used hitherto $\tfrac{1}{2}(q_{-1} + q_1)$ approximates to q_0 and so we have

$$(p_1 - p_{-1})/2h = q_0.$$

The left-hand side is our approximation for p_0', and so we may state the result as a theorem.

THEOREM 5.4

The system of finite difference equations corresponding to the differential equation

$$p(x)y'' + q(x)y' + r(x)y = 0$$

has a symmetric matrix if, and only if

$$p'(x) = q(x).$$

This establishes, from another point of view, the link between self-adjoint equations and symmetric matrices. It is necessary to record another warning. Numerical computation requires careful analysis, and the proof above lacks the rigour which serious numerical analysis needs. Here we are deliberately employing simplified methods in order to give indications of the scope of the subject. For an introduction to more rigorous methods reference may be made to Fox [8] and Fox and Mayers [9].

It was said earlier that the rigorous discussion of the convergence of Fourier series presents many difficulties. The next result shows a way in which the sum of the first N terms of a Fourier series, which has been derived from a function f, approximates to the function. It shows that in a certain respect the Fourier coefficients are the best possible choice. The method employed below is an adaptation of an answer which was given by a schoolboy to a Cambridge scholarship question [15]; his answer gave an illuminating twist to a standard proof.

First of all we will consider a purely geometrical problem in three dimensions (Fig. 5.2). The projections of a point F on the x_1 and x_2 axes are A_1 and A_2 respectively. OA is the vector sum of OA_1 and OA_2. If B is any point in the Ox_1x_2 plane, show that the length FB is minimized by taking B at A. The proof of this is very simple; it is merely a question of recognizing that A is the foot of the perpendicular from F to the plane Ox_1x_2, and so FAB is a right-angled triangle with

$$FB^2 = FA^2 + AB^2. \qquad (5.15)$$

FA is fixed, and so to minimize FB, AB must be made zero.

This is a very simple piece of three-dimensional geometry, but the next problem is to frame exactly the same argument in a space of infinite dimension.

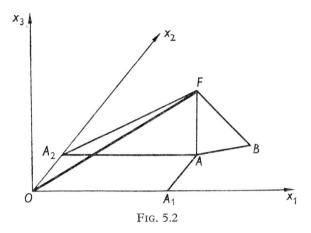

Fig. 5.2

Consider a set of normalized eigenfunctions $\{g_n(x)\}$, which therefore satisfy Eqs. (5.4), or similar equations with the integral over some interval (a, b). Let f be some function which is integrable over the appropriate interval. Then the Fourier coefficients α_n of f are defined by the inner products

$$\alpha_n = [f, g_n].$$

Put

$$f_A = \sum_{n=1}^{N} \alpha_n g_n,$$

that is to say f_A is the sum of the first N terms of the Fourier series of f.

We will now prove the following theorem.

THEOREM 5.5

f_A is the best approximation to f which is obtainable from the first N eigenfunctions in the sense that among all possible choices of numbers β_n ($n = 1, 2, \ldots, N$) the choice

$$\beta_n = \alpha_n, \qquad n = 1, 2, \ldots, N,$$

makes the expression

$$\int_0^1 \left\{ f(x) - \sum_{n=1}^{N} \beta_n g_n(x) \right\}^2 dx$$

a minimum.

Proof

To commence the proof put

$$f_P = f - f_A.$$

f_P thus corresponds to the vector AF in the three-dimensional problem, and we will have to prove that it is perpendicular to all the vectors in the space spanned by $g_1, g_2, \ldots, g_N$. This is easily managed because

$$[f_P, g_k] = [f - f_A, g_k] = \left[f - \sum_{n=1}^{N} \alpha_n g_n,\ g_k \right],$$

$$= [f, g_k] - \sum_{1}^{N} \alpha_n [g_n, g_k],$$

$$= \alpha_k - \alpha_k = 0,$$

using in the first case the definition of α_k and in the second the orthogonal relations in Eqs. (5.4).

Hence f_P is orthogonal to each g_k, and so to any linear combination of them. We continue towards the main result.

Put

$$f_B = \sum_{n=1}^{N} \beta_n g_n.$$

Then

$$\int_0^1 \left\{ f(x) - \sum_{n=1}^{N} \beta_n g_n \right\}^2 dx$$

$$= [(f - f_B), (f - f_B)],$$
$$= [\{(f - f_A) + (f_A - f_B)\}, \{(f - f_A) + (f_A - f_B)\}],$$
$$= [\{f_P + (f_A - f_B)\}, \{f_P + (f_A - f_B)\}],$$
$$= [f_P, f_P] + 2[f_P, (f_A - f_B)] + [(f_A - f_B), (f_A - f_B)].$$

These calculations involve repeated use of the properties of the inner product given in Exercise 5.6. The second bracket is zero as we have shown that f_P is orthogonal to any linear combination of g_ks. Using the orthogonal relations of the g_ks the third bracket reduces to

$$\sum_{n=1}^{N} (\alpha_n - \beta_n)^2.$$

Hence we have

$$\int_0^1 \left\{ f(x) - \sum_{n=1}^N \beta_n g_n \right\}^2 dx = \int_0^1 \left\{ f_P(x) \right\}^2 dx + \sum_{n=1}^N (\alpha_n - \beta_n)^2. \quad (5.16)$$

Equation (5.16) corresponds to Eq. (5.15) in the elementary geometrical proof. Equation (5.16) is a case of Pythagoras's theorem in a space of infinite dimension. The first term on the right-hand side is positive because it is the integral of the square of a real function, and it is independent of the choice of β_n. The second term is positive or zero, and can only be made zero by taking every β_n equal to the corresponding α_n.

This result has been discussed at some length because it shows how the ideas of elementary geometry can be carried over into function spaces. It may be noted that if Fig. 5.2 is used to illustrate the argument then free vectors are being employed, in contrast to the exclusive use of vectors bound to the origin as in Chapter 2. This is done because it gives perhaps the simplest picture in terms of three-dimensional geometry. At the expense of a slightly more complicated figure it could all be done with bound vectors. However, the algebraic proof in no way depends on the geometrical illustration, it is complete on its own.

The ideas of linear spaces and inner products may be used to discuss further the forced oscillations of the string, first considered as Example 3.5. Equation (3.2) in Section 3.3 was

$$(\mathscr{D} - k^2)y = p. \quad (5.17)$$

It will be remembered that the displacement at the point x on the string at time t is $y(x) \sin kt$, the eigenvalues of $\mathscr{D}$ occur for $k^2 = r^2\pi^2 = \lambda_r$, and the corresponding normalized eigenfunctions are

$$y_r(x) = \sqrt{2} \sin r\pi x.$$

The problem is to find how the string responds to the perturbing force $p(x) \sin kt$. Using vector notation for the functions, and proceeding by a method that is quite general, form the inner product of Eq. (5.17) with $\mathbf{y}_r$ (this is projecting along the $\mathbf{y}_r$ axis!). This gives

$$[\mathbf{y}_r, \mathscr{D}\mathbf{y}] - k^2[\mathbf{y}_r, \mathbf{y}] = [\mathbf{y}_r, \mathbf{p}].$$

Using the symmetry of $\mathscr{D}$

$$[\mathscr{D}\mathbf{y}_r, \mathbf{y}] - k^2[\mathbf{y}_r, \mathbf{y}] = [\mathbf{y}_r, \mathbf{p}],$$
$$[\lambda_r \mathbf{y}_r, \mathbf{y}] - k^2[\mathbf{y}_r, \mathbf{y}] = [\mathbf{y}_r, \mathbf{p}],$$

using the eigenvector property of $\mathbf{y}_r$, $\mathscr{D}\mathbf{y}_r = \lambda_r \mathbf{y}_r$. Now put $[\mathbf{y}_r, \mathbf{y}] = a_r$

and $[y_r, p] = p_r$, these are the Fourier coefficients of $y = y(x)$ and $p = p(x)$, then
$$(\lambda_r - k^2)a_r = p_r,$$
so
$$a_r = p_r/(\lambda_r - k^2).$$

This gives the coefficient of y_r in the Fourier expansion of y. Hence

$$y = \sum_r \frac{p_r y_r}{\lambda_r - k^2}. \tag{5.18}$$

Once again we are ignoring problems of convergence. This result is in a general form; in the particular case of the vibrating string the normalized eigenfunctions involved are 1, $\sqrt{2} \sin \pi x$, $\sqrt{2} \sin 2\pi x, \ldots$. With these

$$y(x) = \sqrt{2} \sum_r \frac{p_r}{r^2 \pi^2 - k^2} \sin r\pi x.$$

The physical meaning of this is as follows. The perturbing force $p(x)$ is resolved into its Fourier components, then each Fourier component is weighted in a particular way, being divided by the factor $r^2\pi^2 - k^2$. The terms for which k is near to $r\pi$ are favoured at the expense of the others, and as k tends to $r\pi$ the solution becomes infinite. This is a resonance condition.

The solutions provided by Eq. (5.18), and also the alternative forms of solution given in Method 1 and Method 2 of Section 3.4, are very general as far as the distribution of perturbing forces along the string is concerned, but they are limited to perturbing forces which vary sinusoidally in time. It is of obvious interest to be able to deal with forces whose variation with time is more general than this. Further developments of the theory do this, but we cannot pursue this here.

5.5 Further properties of the Green's function

We will first prove a result about Green's functions that was taken for granted in Chapter 3 because of its physical plausibility.

THEOREM 5.6

The solution of the equation

$$\mathscr{D}y = -(p(x)y')' - q(x)y = f(x), \qquad a \leqslant x \leqslant b \tag{5.19}$$

(*with the usual boundary conditions at a and b*) *for general values of f may be expressed in terms of a function $G(x, \xi)$, which is the solution of the related system of equations.*

5] FURTHER PROPERTIES OF THE GREEN'S FUNCTION

$$\mathscr{D}G = -(pG')' - qG = 0, \qquad a \leqslant x < \xi, \xi < x \leqslant b \quad (5.20)$$

$$[pG']_{\xi-0}^{\xi+0} = -1, \qquad (5.21)$$

with G continuous at $x = \xi$ *and satisfying at a and b the boundary conditions imposed on y.*

$$y(x) = \int_a^b G(x, \xi) f(\xi) \, d\xi.$$

(The reader may need to check that the signs used above are chosen so as to conform with the conventions we have observed hitherto. The types of boundary condition permissible are listed after Eq. (5.11) of Section 5.4.)

Proof

Multiplying Eqs. (5.19) and (5.20) by G and y respectively, and subtracting

$$G(py')' - y(pG')' = -Gf.$$

We next integrate by parts over the range (a, b); but there is the crucial discontinuity in G' at the point ξ, and to overcome this care is necessary. We must integrate over the intervals $(a, \xi - h)$ and $(\xi + h, b)$ separately, and then take the limit as h tends to zero.

Integrating by parts

$$[Gpy']_a^{\xi-h} + [Gpy']_{\xi+h}^b - \int_a^{\xi-h} G' py' \, dx$$

$$- \int_{\xi+h}^b G'py' \, dx - [ypG']_a^{\xi-h} - [ypG']_{\xi+h}^b$$

$$+ \int_a^{\xi-h} pG'y' \, dx + \int_{\xi+h}^b pG'y' \, dx = -\int_a^b Gf \, dx,$$

the limit being taken as $h \to 0$. Using the boundary conditions at a and b, Gpy' vanishes at each; also Gpy' is continuous at ξ and so two more terms cancel as $h \to 0$. This means that the terms arising from the first two sets of square brackets reduce to zero. The integrals on the left-hand side involve precisely the same integrands, and they reduce to zero. There remains

$$[ypG']_a^{\xi-0} + [ypG']_{\xi+0}^b = \int_a^b Gf \, dx.$$

The boundary conditions once more reduce these terms to zero at a and b, leaving

$$-[ypG']_{\xi-0}^{\xi+0} = \int_a^b Gf \, dx.$$

Using Eq. (5.21)
$$y(\xi) = \int_a^b G(x, \xi) f(x) \, dx.$$
Changing the variables
$$y(x) = \int_a^b G(x, \xi) f(\xi) \, d\xi.$$

The last step of all involves the symmetry of the Green's function in x and ξ. This is something which can be observed in all of the examples, but we have given no general proof that the Green's function of a symmetric operator necessarily has this property. The need to establish a proof having been seen, we will rectify the omission.

THEOREM 5.7
The Green's function $G(x, \xi)$ of the differential operator
$$\mathscr{D} = -\frac{d}{dx}\left\{p(x)\frac{d}{dx}\right\} - q(x)$$
is symmetric in x and ξ.

Proof
Using the equations employed in the proof of the previous theorem, let $u(x)$ and $v(x)$ be solutions valid over the interval (a, b), and such that $u(x)$ satisfies the required boundary condition at a and $v(x)$ satisfies the boundary condition at b. (In general only eigenfunctions satisfy both conditions simultaneously, and our concern at the moment is not with eigenfunctions.)

Then
$$(pu')' + qu = 0, \quad a \leqslant x \leqslant b,$$
and
$$(pv')' + qv = 0, \quad a \leqslant x \leqslant b.$$

Multiplying these equations by v and u respectively, and subtracting
$$v(pu')' - u(pv')' = 0.$$
Taking the indefinite integral by parts,
$$vpu' - \int v'pu' \, dx - upv' + \int v'pu' \, dx = \text{constant} = c \text{ (say)}.$$
Hence
$$p(vu' - uv') = c. \tag{5.22}$$

We may now construct $G(x, \xi)$ explicitly, putting
$$G(x, \xi) = u(x) v(\xi)/c, \quad a \leqslant x \leqslant \xi,$$
$$= u(\xi) v(x)/c. \quad \xi \leqslant x \leqslant b.$$

It may now be checked that G has all the required properties. In particular the property

$$[pG']_{\xi-0}^{\xi+0} = -1$$

arises from Eq. (5.22). The form of the expressions for $G(x, \xi)$ shows that it is symmetric in x and ξ.

Exercise 5.8 (harder)
  In Exercise 5.2 the flexibility matrix **G** was expressed in terms of the eigenvectors by the formula

$$\mathbf{G} = \sum \frac{1}{\lambda_r} \mathbf{x}_r \mathbf{x}_r',$$

$\mathbf{x}_r$ being the eigenvector of the stiffness matrix **A** with eigenvalue λ_r. (Incidentally, of course, the eigenvectors of **G** are just the same as the eigenvectors of **A**.)
  Show that with differential operators the Green's function may be expected to have the expansion

$$G(x, \xi) = \sum_r \frac{y_r(x) \, y_r(\xi)}{\lambda_r}.$$

Whilst this result is frequently true there can be difficulties with convergence, and the formula does not always hold.

Exercise 5.9 (harder)
  The Green's function for dynamic systems $\Gamma(x, \xi)$ was described in Section 3.4. Leaving aside problems of convergence, show that (tentatively)

$$\Gamma(x, \xi) = \sum_r \frac{y_r(x) \, y_r(\xi)}{\lambda_r - k^2}.$$

The solution to

$$(\mathscr{D} - k^2)y = p,$$

where $\mathscr{D}$ is a symmetric differential operator, in terms of the Green's function is

$$y(x) = \int_a^b \Gamma(x, \xi) \, p(\xi) \, d\xi.$$

Show that this solution is equivalent to the solution in Eq. (5.18). Observe incidentally that as $k \to 0$, $\Gamma \to G$.

5.6 Integral operators

Our work has been concerned mainly with two types of operator on linear spaces—matrices operating on vectors in spaces of finite dimension, and differential operators on functions in spaces of infinite dimension. In passing, some work was done with difference operators which operated on spaces of sequences. These have not received any attention in the present chapter, but it is possible to adapt the methods and results of this chapter to difference operators quite easily.

The Green's function introduces another type of operator as it enables Sturm–Liouville systems of equations to be reformulated as integral equations. Thus in Section 3.4 the equation

$$(\mathscr{D} - k^2 y) = p$$

was converted to

$$y = \mathscr{G}(p + k^2 y), \tag{5.23}$$

where $\mathscr{G}$ was the integral operator which arose as the inverse of the differential operator $\mathscr{D}$. There are particular cases of this in Examples 3.5, 3.6 and 4.1; other cases may clearly be treated in a similar way. It will be remembered that $\mathscr{G}$ is defined by

$$\mathscr{G}f = \int G(x, \xi) f(\xi) \, d\xi,$$

the integral being taken over the appropriate range.

Equation (5.23) written out more fully is

$$y(x) = \int G(x, \xi)[p(\xi) + k^2 y(\xi)] \, d\xi,$$

and this may be regarded as an equation for y; and because one of the appearances of y is under the integral sign it is called an *integral equation*.

The eigenvalues and eigenfunctions associated with the operator $\mathscr{D}$ are defined by the equation

$$\mathscr{D}y = \lambda y. \tag{5.24}$$

The corresponding integral equation is

$$y = \lambda \mathscr{G}y. \tag{5.25}$$

The term λ now appears on the other side of the equation, just as the matrix equation $\mathbf{Ax} = \lambda \mathbf{x}$ converts to $\mathbf{x} = \lambda \mathbf{Gx}$ when $\mathbf{G}$ is the inverse of $\mathbf{A}$.

Given the function $G(x, \xi)$, Eq. (5.25) can be regarded as posing the eigenvalue problem—are there any values λ, and any functions $y(x)$ satisfying Eq. (5.25)? Some writers take this as the standard form of equation, and it is sometimes perplexing to inexperienced readers that the eigenvalue question appears to be posed in inconsistent ways in Eqs. (5.24) and (5.25), but this is brought about by the reciprocal nature of $\mathscr{D}$ and $\mathscr{G}$. $\mathscr{D}$ and $\mathscr{G}$ arise from the same physical problems, and if the standard equations are taken as (5.24) and (5.25) then they have the same eigenvalues. If the problem is posed in the form of finding the eigenvalues of $\mathscr{G}$ according to $\mathscr{G}y = \lambda y$, then the eigenvalues of $\mathscr{G}$ are the reciprocals of the eigenvalues of $\mathscr{D}$.

Our approach was based on physical problems which led to differential equations. We are now seeing how these differential equations may be translated into integral equations, and in these integral equations Green's function plays a key role. Green's function has been seen not only in an abstract form, as the solution of a particular system of equations, it was introduced from physically plausible considerations and it is always possible to call a clear physical picture to mind. $G(x, \xi)$ gives the displacement at the point x when a unit force is applied at the point ξ.

The beginner may wonder what advantages the formulation of physical problems in terms of integral equations has over the more familiar formulations in terms of differential equations. It could be argued that the use of differential equations is only a matter of habit, and that integral relations give a better expression of the underlying experimentally available phenomena. This might reasonably be said in electrical and gravitational theory where a given body experiences a force or has a potential which is the sum total of the effects of all the other charges or masses in the region under discussion. That is to say, an integral often describes something directly measurable, whereas it might be argued that a quantity like $\partial^2 V/\partial r^2$ is more of an abstraction, and is not available to direct experimental measurement in the same way.

The formulation of integral equations instead of differential equations does not give any computational advantages at an elementary level; indeed text books contain a number of well-tried algorithms for differential equations when there are no corresponding ones for integral equations. But when one is concerned with numerical work, numerical integration is a far more accurate and reliable process than numerical differentiation; and when a comprehensive theory is required any function that can be differentiated can be integrated, but not vice versa. Integration tends to make functions 'better' in the sense of being more amenable to analytic treatment, whereas differentiation can make them worse.

Furthermore, the theory of multiple integrals is more like the theory of single integrals than the theory of partial differential equations is like the theory of ordinary differential equations. This helps in constructing theories in which there is more than one space variable.

Again, the integral operator $\mathscr{G}$ corresponding to the differential operator $\mathscr{D}$ has the advantage of involving a single integration, whereas $\mathscr{D}$ involves two differentiations. Example 4.2 showed how a fourth order differential equation can be replaced by an equation involving one integration.

The physical properties of Green's function suggest other possibilities. Given a complicated physical system which is linear in its behaviour, it might be possible to obtain information about the Green's function experimentally, even if it is too complicated to formulate theoretical equations. From a table of numerical values of the Green's function other properties, such as the frequencies of the normal modes of vibration, might be estimated.

Finally, some purely mathematical questions arise. G need not start out as the Green's function associated with some previously known differential operator; it might be any symmetric function of two variables. We may consider Eq. (5.25) and ask if it has eigenvalues and eigenfunctions. This is the starting point for an independent theory of integral equations, and from a theoretical point of view it is a good place to begin. However, when beginning here it is not easy to give convincing, small scale numerical exercises for the student and he is left wondering how the eigenvalues and eigenfunctions are actually calculated. A short, and perhaps over-simplified, answer is to say that Rayleigh's principle can be applied and it leads to methods of calculating the eigenvalues successively, to any required accuracy, if sufficient computing power is available [7,13]. Some of the methods in the next chapter also apply.

It turns out here, as in so many other branches of mathematics, that complex numbers give a much more nicely rounded theory than real numbers alone, and also that complex numbers are needed when the methods of working which we have developed in the last three chapters are applied to the problems of quantum mechanics. This leads to the study of Hilbert space, a branch of pure mathematics that has been much developed since it was established towards the beginning of the century. Hilbert space is the space of complex valued functions which satisfy certain conditions of integrability, and experience has shown it to be a convenient space in which to formulate theoretical questions on symmetric operators and their related eigenfunction expansions.

References

1. BERBERIAN, S. K., *Introduction to Hilbert Space*, OUP (1961).
2. BOURBAKI, N., *Elements d'histoire des Mathématiques*, Hermann, Paris (1960).
3. BROWN, A. L. and PAGE, A., *Elements of Functional Analysis*, Van Nostrand Reinhold, London (1970).
4. BURKILL, J. C., *The Theory of Ordinary Differential Equations*, Oliver and Boyd (1962).
5. CHURCHILL, R. V., *Fourier Series and Boundary Value Problems*, McGraw-Hill (1941).
6. COPPEL, W. A., 'J. B. Fourier—on the occasion of his two-hundredth birthday', *Am. math. Mon.*, 76, 5, pp. 468–83 (May 1969).

7. COURANT, R. and HILBERT, D., *Methods of Mathematical Physics*, 2 vols., Interscience, New York (1953).
8. FOX, L., *The Numerical Solution of two-point Boundary Problems*, OUP (1957).
9. FOX, L. and MAYERS, D. F., *Computing Methods for Scientists and Engineers*, OUP (1968).
10. INCE, E. L., *Ordinary Differential Equations*, Longmans (1927).
11. LANCZOS, C., *Applied Analysis*, Pitman (1957).
12. LANCZOS, C., *Linear Differential Operators*, Van Nostrand (1961).
13. MIKHLIN, S. G., *Integral Equations*, Pergamon (1957).
14. WILKINSON, J. H., *The Algebraic Eigenvalue Problem*, OUP (1965).
15. YOUD, N., 'An original solution of a problem in Calculus', *Mathematical Spectrum*, **3**, 1970/71, 1, pp. 17–21.

6
Iterated Powers of Matrices

Various naturally occurring problems of growth and development bring out the need to take powers of a matrix, or to generate a sequence of vectors by multiplying successively by a matrix. Investigation shows that this leads to practical methods of finding the eigenvalues and eigenvectors of the matrix which are frequently more convenient than using determinants and the characteristic equation. The final section of this chapter takes a brief look at certain continuous processes in order to show that matrix methods are not limited to situations in which growth is by discrete steps.

Those interested in the applications of matrices will meet stochastic matrices and Markov chains. Those whose interest is more theoretical will find some further methods of computing eigenvalues and eigenvectors.

6.1 A square root process

EXAMPLE 6.1

The following method for extracting square roots was described by Theon of Smyrna in the second century A.D. If we tabulate values of (x, y), starting with $(1, 1)$ and at each step replacing (x, y) by $(x + 2y, x + y)$ we get

x	y	x^2	y^2
1	1	1	1
3	2	9	4
7	5	49	25
17	12	289	144
41	29	1681	841
99	70	9801	4900, etc.

From the table it appears that x^2/y^2 gets closer and closer to 2, and hence that x/y gives a sequence of approximations to $\sqrt{2}$. It is not difficult to prove that this is so. If the new values of x and y are denoted by x' and y', then

$$\frac{x'}{y'} - \sqrt{2} = \frac{x+2y}{x+y} - \sqrt{2},$$

$$= \frac{x+2y - \sqrt{2}x - \sqrt{2}y}{x+y},$$

$$= \frac{(\sqrt{2}-1)(\sqrt{2}y - x)}{x+y},$$

$$= \frac{\sqrt{2}-1}{(x/y)+1}\left(\sqrt{2} - \frac{x}{y}\right).$$

But $\sqrt{2} - 1 < 1$, and $x/y \geqslant 1$ at every stage, and so

$$\left|\frac{x'}{y'} - \sqrt{2}\right| < \frac{1}{2}\left|\frac{x}{y} - \sqrt{2}\right|.$$

So we see that at each stage the error is at least halved, and so by continuing far enough $\sqrt{2}$ may be approximated to any required accuracy. We see also that the approximations are alternately too big and too small.

Exercise 6.1
Investigate the effect of taking other numbers as the initial values of x and y. Adapt the method to find the square roots of numbers other than 2.

The method can be written in terms of matrices as follows. The transformation involved is

$$(x, y) \rightarrow (x', y') = (x, y)\mathbf{A},$$

where
$$\mathbf{A} = \begin{bmatrix} 1 & 1 \\ 2 & 1 \end{bmatrix}.$$

The sequence of values of x and y is given by

$$(x, y), (x, y)\mathbf{A}, (x, y)\mathbf{A}^2, (x, y)\mathbf{A}^3, \ldots .$$

This means that $\sqrt{2}$ is somehow connected with powers of the matrix $\mathbf{A}$. Before pursuing the nature of this connexion we will consider some other situations in which it is of value to take powers of matrices.

6.2 Beetles

EXAMPLE 6.2
A certain species of beetle behaves in the following way [3, 12].
Half the beetles which are born, on the average, survive their first birthday and live into a second year.

Of these, one-third survive their second birthday and live into a third year.

By the end of the third year they are all dead.

During the third year of life each beetle produces, on the average, six offspring.

If we start with a population of 3000 beetles, with 1000 in each age group, how may we expect the population to fluctuate in the future?

Direct calculation, applying the information to the numbers in the age groups each year, gives:

Starting situation, working in thousands (1, 1, 1), 3000 in all;
After 1 year (6, $\frac{1}{2}$, $\frac{1}{3}$), 6833 in all;
After 2 years (2, 3, $\frac{1}{6}$), 5166 in all;
After 3 years (1, 1, 1), 3000 in all.
The cycle then repeats.

After a little investigation we can see that each row vector is produced from the one before by postmultiplying it by the matrix

$$\mathbf{A} = \begin{bmatrix} 0 & \frac{1}{2} & 0 \\ 0 & 0 & \frac{1}{3} \\ 6 & 0 & 0 \end{bmatrix}.$$

Thus

$$(1,\ 1,\ 1)\mathbf{A} = (6,\ \tfrac{1}{2},\ \tfrac{1}{3}),\quad \text{etc.}$$

What is the significance of the entries in the matrix?

It is easy to see that the entry $\frac{1}{2}$ in row 1, column 2, expresses that one-half of the beetles survive from the first year group and move into the second year group; and the $\frac{1}{3}$ in row 2, column 3, expresses that one-third of the second year group survive into the third year group. Put this way the general pattern emerges and it is easy to see how to apply the method to a population which is classified into more than three age groups. Terms denoting survival rates appear in the diagonal of the matrix situated immediately above the principal diagonal.

The 6, corresponding to the average six offspring from each of the beetles in their third year of life, is situated in row 3, column 1. There is, as it were, a transfer of numbers from the third year group to the first year group, but of course it is a different set of individuals that is involved. More generally, if we are dealing with a population in which births can occur in any year of life, then terms indicating the birth rates in the different year groups are located in the first column of the matrix.

It is interesting to reconsider the previous problem, but with a different birth pattern. If each beetle, on the average, produces one more in its second year of life, and three more in its third year of life the matrix is

$$\mathbf{A} = \begin{bmatrix} 0 & \frac{1}{2} & 0 \\ 1 & 0 & \frac{1}{3} \\ 3 & 0 & 0 \end{bmatrix}.$$

With the same initial conditions this time the sequence of vectors is

$(1, 1, 1),$ 3000 in all;
$(4, \frac{1}{2}, \frac{1}{3}),$ 4833 in all;
$(\frac{3}{2}, 2, \frac{1}{6}),$ 3666 in all;
$(\frac{5}{2}, \frac{3}{4}, \frac{2}{3}),$ 3917 in all;
$(\frac{11}{4}, \frac{5}{4}, \frac{1}{4}),$ 4250 in all; etc.

This situation is more complicated. In fact the population is tending to a limiting value of 4000, distributed into age groups in the ratio 2·4:1·2:0·4. It is not obvious how to prove this, using only the techniques which we have at this moment; but it can be proved by methods to be discussed in Section 6.6. However, it is easy enough to verify that a population of (2400, 1200, 400) is stable. This is, in fact, a row eigenvector of the matrix with eigenvalue unity. Column eigenvectors were introduced in Section 2.5; row eigenvectors may be defined as vectors satisfying a relation

$$\mathbf{yA} = \lambda \mathbf{y},$$

with λ a scalar, and they have properties analogous to those of column eigenvectors.

Two further cases may be investigated by the reader. In the matrix

$$\begin{bmatrix} 0 & \frac{1}{3} & 0 \\ 6 & 0 & \frac{1}{2} \\ 24 & 0 & 0 \end{bmatrix}$$

the birth rates look relatively high, and it will be found that with this matrix the population of beetles increases without bound.

In the case

$$\begin{bmatrix} 0 & \frac{1}{16} & 0 \\ 3 & 0 & \frac{1}{12} \\ 6 & 0 & 0 \end{bmatrix}$$

the death rates are relatively high, and it will be found that the beetles are doomed to extinction.

This is a suitable place at which to state a striking result. Whether or not the population expands, stays constant or contracts is determined by the largest eigenvalue of the matrix; if this is less than one it contracts, if this is greater than one the population expands. This result will be proved later, in Section 6.6; but it is an easy enough result to understand, and it is remarkable that however complicated the pattern of survival and birth rates of a population this simple index decides whether it will increase or decline.

6.3 Chemical examples

EXAMPLE 6.3

The same kind of analysis can be applied to chemical reactions. Hydrogen and oxygen react according to the formula

$$2H_2 + O_2 \to 2H_2O.$$

Chemists now believe that molecules H_2 and O_2 will not react this way on their own; the reaction is attributed to the presence of radicals H, O and OH, which participate in three simultaneous reactions

$$O + H_2 \longrightarrow OH + H,$$
$$OH + H_2 \longrightarrow H_2O + H,$$
$$H + O_2 \longrightarrow OH + O.$$

(For our purposes it is not necessary to indicate the electrical charges on the radicals.)

A single radical starts a chain reaction according to the pattern shown in Fig. 6.1. The uranium nuclear fission reaction is a somewhat similar branched chain reaction where the propagating particles are neutrons.

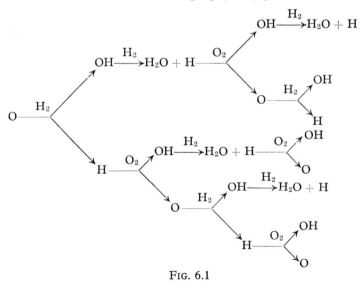

Fig. 6.1

For mathematical simplicity we will assume that all three reactions take place at the same speed, and that there is an unlimited supply of oxygen and hydrogen radicals. At time t_0 we will start a chain reaction with one oxygen radical, and calculate how many O, OH and H's there are at each successive stage.

Denote the number of O, OH and H radicals at stage n by o_n, $(oh)_n$ and h_n respectively.

Then
$$o_{n+1} = h_n,$$
$$(oh)_{n+1} = o_n + h_n, \quad \text{with } o_0 = 1, (oh)_0 = 0, h_0 = 0, \quad (6.1)$$
and
$$h_{n+1} = o_n + (oh)_n.$$

This is a set of simultaneous difference equations, which may be tackled with techniques of varying sophistication. We consider first how the problem might be solved without matrices.

Method 1

$$o_{n+1} = h_n = o_{n-1} + (oh)_{n-1} = o_{n-1} + o_{n-2} + h_{n-2},$$
and so
$$o_{n+1} = 2o_{n-1} + o_{n-2}. \quad (6.2)$$

This is a third order recurrence relation, and the standard technique for solving it is to form the auxiliary equation

$$x^3 - 2x - 1 = 0,$$
$$(x + 1)(x^2 - x - 1) = 0, \quad (6.3)$$

with solutions
$$x = -1, \quad \tfrac{1}{2}(1 \pm \sqrt{5}).$$

(The reader who is unfamiliar with this method will find it described in most first year University texts on traditional algebra.)

This leads to the general solution

$$o_n = (-1)^n p + q[\tfrac{1}{2}(1 + \sqrt{5})]^n + r[\tfrac{1}{2}(1 - \sqrt{5})]^n,$$

where p, q and r have to be chosen to fit the initial conditions, which are

$$o_0 = 1, \quad o_1 = 0, \quad \text{and} \quad o_2 = 1.$$

The numerical values of p, q and r may be found as an exercise, but they are of no immediate interest.

We can likewise seek the recurrence relation for h_n.

$$h_{n+1} = o_n + (oh)_n = h_{n-1} + o_{n-1} + h_{n-1},$$
$$h_{n+1} = 2h_{n-1} + h_{n-2}. \quad (6.4)$$

It may be a surprise that Eqs. (6.2) and (6.4) are the same, and at the moment we have no theory to account for this.

What is the recurrence relation for $(oh)_n$? A little manipulation will show once more that

$$(oh)_{n+1} = 2(oh)_{n-1} + (oh)_{n-2}. \tag{6.5}$$

Equations (6.4) and (6.5) may now be solved as (6.2) was, but the final solutions will differ as the initial conditions differ. This situation could be developed further, but we will merely note that the equation (6.3) clearly has some special connexion with system (6.1). Let us leave it at this and employ matrix methods.

Method 2

Equation (6.1) may be written in matrix form. If

$$\mathbf{q}_n = (o_n, (oh)_n, h_n) \quad \text{and} \quad \mathbf{A} = \begin{bmatrix} 0 & 1 & 1 \\ 0 & 0 & 1 \\ 1 & 1 & 0 \end{bmatrix}$$

then

$$\mathbf{q}_{n+1} = \mathbf{q}_n \mathbf{A},$$

and in general

$$\mathbf{q}_n = \mathbf{q}_0 \mathbf{A}^n.$$

This provides a formula for the number of radicals present at successive stages of the chemical reaction. Conceptually the formula is very useful because it shows that we may think of the successive stages of the reaction as being described by a geometrical progression of vectors

$$\mathbf{q}_0, \ \mathbf{q}_0 \mathbf{A}, \ \mathbf{q}_0 \mathbf{A}^2, \ \mathbf{q}_0 \mathbf{A}^3, \dots,$$

If we wish to have some easy way of calculating the general term of this sequence, without calculating all of the previous terms successively, further techniques are called for. These will be discussed later in this chapter in Section 6.6. For the moment we wish merely to investigate a number of situations, all of which lead to the same problem—the evaluation of powers of a matrix. When we have seen a number of problems to which this process is the key it will be worthwhile to develop techniques. We may observe in passing that the equation

$$x^3 - 2x - 1 = 0,$$

which occurred repeatedly in Method 1, is the characteristic equation of matrix $\mathbf{A}$.

A simpler chemical example of the same kind is provided by the reaction between hydrogen and chlorine, normally written as

$$H_2 + Cl_2 \rightarrow 2HCl.$$

This is now considered to involve two simultaneous reactions involving radicals,
$$H + Cl_2 \to HCl + Cl$$
$$Cl + H_2 \to HCl + H.$$

Exercise 6.2

Formulate this in matrix terms for yourself. Carry out some stages of the iteration with various starting conditions.

6.4 The educational system

Similar methods of analysis have been applied to the educational system in Great Britain. Redfern [16] gives Table 6.1 showing the movement of students through primary and secondary education, then through further education, the universities and colleges of education into teaching posts. These teaching posts are, in their turn, at the different levels of education. Movement of teachers takes place from one branch to another, and of course as the process goes on some students and teachers move out of the educational system altogether. Redfern's figures are for the year 1962–3. Table 6.2 is derived from this data, giving a 10×10 matrix showing the rates of transfer between the various categories. The table is to be interpreted in the following way. If we ask what became of the students who started the year in secondary schools, we see that 0·802 of them were still at secondary school in the following autumn, 0·019 had moved on to further education, 0·007 to university, 0·004 to colleges of education and 0·161 had moved outside education. This last figure does not include the 0·007 who had died or emigrated.

Tabulating the data in this form it is necessary to account for new births, who will be included in the 'outside education' category until they are of school age, and immigrants. If we then start in the autumn of the year, and y_0 is a row vector describing the actual number of persons in the different categories at the starting time, and x_0 is the row vector describing births and immigration during the first year, then the numbers in the different categories as the second year begins are given by the new vector

$$y_1 = y_0 P + x_0, \tag{6.6}$$

where P is the matrix in Table 6.2.

If it is reasonable to assume that P will remain constant, at least for a few years, then this relation may be used to estimate trends, using the best available estimates for x_1, x_2, etc. as the process goes on.

The Department of Education and Science is designing a computer model of the educational system for actual use. At the time of writing its

TABLE 6.1 *Simple education/manpower flow matrix for England and Wales. Period covered: autumn 1962 to autumn 1963 (Numbers in thousands)*

		Students to:					Teachers to:				Outside education (10)	Deaths and emigration (11)	Total autumn 1962 (12)
		Primary (1)	Secondary (2)	Further education (3)	University (4)	College of education (5)	Primary and secondary (6)	Further education (7)	University (8)	College of education (9)			
Students from:													
Primary	(1)	3767	640	—	—	—	—	—	—	—	—	33	4,440
Secondary	(2)	—	2552	60	22	13	—	—	—	—	512	21	3,180
Further education	(3)	—	—	101	1	—	1	—	—	—	49	5	157
University	(4)	—	—	—	75	—	4	—	1	—	16	2	98
College of education	(5)	—	—	—	—	32	14	1	—	—	1	—	48
Teachers from:													
Primary and secondary	(6)	—	—	—	—	—	293	1	—	1	24	1	320
Further education	(7)	—	—	—	—	—	1	26	—	—	1	—	28
University	(8)	—	—	—	—	—	—	—	12	—	1	—	13
College of education	(9)	—	—	—	—	—	—	3	—	4	—	—	4
Outside education	(10)	712	—	8	4	8	12	3	1	—	37,072	864	38,684
Birth and immigration	(11)	30	20	7	2	1	—	—	—	—	1,250	—	1,310
Total at autumn 1963	(12)	4509	3212	176	104	54	325	31	14	5	38,926	926	48,282

Note: Only full-time students and teachers are shown. Figures are illustrative.

TABLE 6.2 *Transition proportions matrix for England and Wales. Period covered: autumn 1962 to autumn 1963*

		Students to:					Teachers to:					
		Primary (1)	Secondary (2)	Further education (3)	University (4)	College of Education (5)	Primary and secondary (6)	Further education (7)	University (8)	College of education (9)	Outside education (10)	Deaths and emigration (11)	Total (12)
Students from:													
Primary	(1)	0.849	0.144	—	—	—	—	—	—	—	—	0.007	1.000
Secondary	(2)	—	0.802	0.019	0.007	0.004	—	—	—	—	0.161	0.007	1.000
Further education	(3)	—	—	0.64	0.01	—	0.01	—	—	—	0.31	0.03	1.00
University	(4)	—	—	—	0.77	—	0.04	—	0.01	—	0.16	0.02	1.00
College of education	(5)	—	—	—	—	0.67	0.29	0.02	—	—	0.02	—	1.00
Teachers from:													
Primary and secondary	(6)	—	—	—	—	—	0.918	0.003	—	0.001	0.075	0.003	1.000
Further education	(7)	—	—	—	—	—	0.02	0.94	—	—	0.04	—	1.00
University	(8)	—	—	—	—	—	—	—	0.94	—	0.06	—	1.00
College of education	(9)	—	—	—	—	—	—	—	—	0.95	0.05	—	1.00
Outside education	(10)	0.018	—	0.0002	0.0001	0.0002	0.0003	0.0001	—	—	0.959	0.022	1.000

Note: Only full-time students and teachers are shown.
Source: P. Redfern. *Input–output analysis and its application to education and manpower planning*, HMSO, 1967.

final form has still to be decided; but the system just described is one possible form.

6.5 Stochastic matrices

Some of the previous examples involve situations where matrix elements denoted the probability of a transition from one state to another. There was a certain probability that a two-year-old beetle would survive to become a three-year-old one, and this probability was entered at the appropriate place in a matrix. Matrix multiplication is closely related to a natural way of combining probabilities, as we will make clear by means of further examples. The first example does not involve probabilities at all, but it leads on to other situations which do.

EXAMPLE 6.4

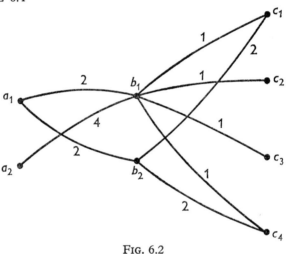

FIG. 6.2

Figure 6.2 shows a schematic map of the interconnexions between the airports in three different countries a, b and c. The figures beside the links denote the number of choices along the links, i.e. a 4 indicates that four airlines fly services along that route. The information may be tabulated thus:

$$\mathbf{P}: \quad \begin{array}{c|cc} & b_1 & b_2 \\ \hline a_1 & 2 & 2 \\ a_2 & 4 & 0 \end{array} \qquad \mathbf{Q}: \quad \begin{array}{c|cccc} & c_1 & c_2 & c_3 & c_4 \\ \hline b_1 & 1 & 1 & 1 & 1 \\ b_2 & 2 & 0 & 0 & 2. \end{array}$$

What table gives the number of choices of route between the airports in country a and country c? A route from a_1 to c_1 may pass through b_1 or through b_2. In the first case there are 2×1 choices and in the second case 2×2 choices. This means 6 choices in all, and the other entries for the table may be worked out in the same way, giving

	c_1	c_2	c_3	c_4
a_1	6	2	2	6
a_2	4	4	4	4

This is simply the familiar matrix product **PQ**. The map in Fig. 6.2, and similar maps, presents a very fertile teaching situation as many different kinds of matrix product arise corresponding to various questions which may be asked about the map. There is no space in which to investigate these questions here, but they are discussed in References [8, 9].

EXAMPLE 6.5

A somewhat artificial situation will serve to introduce the next idea. Bill and Ted play a peculiar gambling game. The rules are

(i) you toss a penny;
(ii) if you throw a head then you draw a card from an ordinary pack, if you throw a tail then you draw a card from among the *black* cards only.

If Bill has an ordinary penny and Ted has a penny with two heads what are their respective probabilities of drawing a club, diamond, heart or spade? This must be compared carefully with the previous example.

We may tabulate:

	Head	Tail
Bill	$\frac{1}{2}$	$\frac{1}{2}$
Ted	1	0

and

	C	D	H	S
Head	$\frac{1}{4}$	$\frac{1}{4}$	$\frac{1}{4}$	$\frac{1}{4}$
Tail	$\frac{1}{2}$	0	0	$\frac{1}{2}$

and we find quite easily that

	C	D	H	S
Bill	$\frac{3}{8}$	$\frac{1}{8}$	$\frac{1}{8}$	$\frac{3}{8}$
Ted	$\frac{1}{4}$	$\frac{1}{4}$	$\frac{1}{4}$	$\frac{1}{4}$

Figure 6.2. still applies, with slight modifications, but gives a rather more abstract representation of the situation than before. We see further that the matrices have to be combined as before, using the fractions instead of the previous whole numbers. Now we are multiplying two matrices of *transition probabilities* to get a third matrix of transition probabilities.

The element in position *rs* of such a matrix gives the probability that the system will move from state *r* to state *s*. Since it must move to one of the various alternative states the sum of the elements in any row is unity, but the same is not true, in general, for the columns.

We may continue to introduce further ideas by means of other examples.

EXAMPLE 6.6

Ann, Bernard, Claire and Douglas are throwing a ball to one another.
Ann always throws it to Douglas;
Bernard is equally likely to throw it to anybody else;
Claire throws it to the boys, with equal frequency;
Douglas throws it to Claire twice as often as to Ann, and never throws it to Bernard.

The situation is described by the matrix of transition probabilities.

$$\mathbf{M}: \quad \begin{array}{c|cccc} & A & B & C & D \\ \hline A & 0 & 0 & 0 & 1 \\ B & \frac{1}{3} & 0 & \frac{1}{3} & \frac{1}{3} \\ C & 0 & \frac{1}{2} & 0 & \frac{1}{2} \\ D & \frac{1}{3} & 0 & \frac{2}{3} & 0 \end{array}.$$

In this case there are four states. In state 1, the ball is with Ann, etc.; and the entries in the matrix show the probabilities of the ball going from one child to another.

What is the probability that the ball will go from

Bernard to Claire in two throws?
Claire to Ann in three throws?   etc.

These questions may also be answered by taking powers of the matrix; the technique being merely an extension of the methods of the previous problems.

The terms in the *n*th power of the matrix give the probabilities of a transition taking place in *n* moves.

This situation is an example of an *ergodic chain*, or *Markov chain with non-absorbing states*. An *absorbing state* is one in which nothing further happens—somebody who keeps hold of the ball. Absorbing states are slightly harder to handle, but they lead to examples which are rather more interesting. It is a useful trick to regard the game as continuing, even though an absorbing state has been reached, by adopting the fiction that the person who holds on to the ball continues to throw it to himself.

The matrix **M** is an example of a *stochastic* matrix—a square matrix in which the sum of the elements in each row is one.

Exercise 6.3
From the statistical interpretation of these matrices we can see that the row sums in higher powers continue to be one. Can you show this as an *algebraic* necessity?

We now go on to the rather more interesting examples in which there are absorbing states.

EXAMPLE 6.7
I invite you to play a gambling game with me. We put down bets and then we toss a die in turn. If you throw a 6 you win outright. If you throw 4 or 5 you have another go. If you throw 1, 2 or 3 you pass the die to the other player. By the way, I want to start. Of course it isn't fair! But how much advantage does the start give me?

There are four states, *My go*, *Your go*, *I win*, *You win*; and there are, as before, probabilities of transition between the states. The winning states bring the game to a conclusion. We intend to attack the problem by iterating powers of a matrix, and so to handle the *stop* situation whilst iterating we adopt the convention that any game continues for an indefinite number of moves once a win for either player has been reached.

The matrix of transition probabilities is

	My go	Your go	I win	You win
My go	$\frac{1}{3}$	$\frac{1}{2}$	$\frac{1}{6}$	0
Your go	$\frac{1}{2}$	$\frac{1}{3}$	0	$\frac{1}{6}$
I win	0	0	1	0
You win	0	0	0	1

$= \mathbf{A}$ (say).

Once again $\mathbf{A}^2$ contains the probabilities of transitions between the states occurring in two stages, $\mathbf{A}^3$ in three stages and so on. It is interesting, by repeatedly squaring, to compute $\mathbf{A}^2$, $\mathbf{A}^4$, etc., to see how the situation develops. Thus $\mathbf{A}^4$ gives

	My go	Your go	I win	You win
My go	313	312	148	138
Your go	312	313	138	148
I win	0	0	1296	0
You win	0	0	0	1296

$\div 1296$.

Note that all the matrix entries have to be divided by 1296. Note also how the device of using a unit element in the diagonal positions corresponding to the absorbing states ensures that the element 148/1296 in the (My go, I win) position gives the probability of my winning in 4 moves *or less* when I start. Likewise the element 138/1296 in the (My go, You win) position gives the probability of your winning in 4 moves *or less* when I start.

To estimate our respective chances it is necessary to consider the behaviour of $\mathbf{A}^n$ as $n \to \infty$. We can see that some general theory is needed if only to reduce the arithmetic. This example will be continued when we have more theory at our disposal.

We have not so far discussed the multiplication of *block matrices*. The idea is simple, but it is one of the reasons why matrices are such powerful numerical tools. A particular case may make the general method clear.

Given the following matrices

$$\begin{bmatrix} a & b & c \\ d & e & f \\ g & h & j \end{bmatrix} \qquad \begin{bmatrix} r & s & t \\ u & v & w \\ x & y & z \end{bmatrix}$$

these may be multiplied at length in the usual way. They may also be partitioned into *blocks*, thus

$$\left[\begin{array}{cc|c} a & b & c \\ d & e & f \\ \hline g & h & j \end{array}\right] \qquad \left[\begin{array}{cc|c} r & s & t \\ u & v & w \\ \hline x & y & z \end{array}\right].$$

We may write these as

$$\left[\begin{array}{c|c} \mathbf{A} & \mathbf{C} \\ \hline \mathbf{G} & \mathbf{J} \end{array}\right] \qquad \left[\begin{array}{c|c} \mathbf{R} & \mathbf{T} \\ \hline \mathbf{X} & \mathbf{Z} \end{array}\right]$$

where the capital letters denote the blocks or *sub-matrices* in an obvious way.

Now, as an experiment, multiply out these matrices by the ordinary rules as if they were 2×2 matrices with the capital letters as individual terms. This gives

$$\begin{bmatrix} \mathbf{AR} + \mathbf{CX} & \mathbf{AT} + \mathbf{CZ} \\ \mathbf{GR} + \mathbf{JX} & \mathbf{GT} + \mathbf{JZ} \end{bmatrix}.$$

It is essential to preserve the correct order, since $\mathbf{AR} \neq \mathbf{RA}$, etc. Now

$$\mathbf{AR} + \mathbf{CX} = \begin{bmatrix} a & b \\ d & e \end{bmatrix} \begin{bmatrix} r & s \\ u & v \end{bmatrix} + \begin{bmatrix} c \\ f \end{bmatrix} [x \quad y],$$

$$= \begin{bmatrix} ar + bu & as + bv \\ dr + eu & ds + ev \end{bmatrix} + \begin{bmatrix} cx & cy \\ fx & fy \end{bmatrix},$$

$$= \begin{bmatrix} ar + bu + cx & as + bv + cy \\ dr + eu + fx & ds + ev + fy \end{bmatrix}.$$

This is the top left-hand block of terms in the product of the original 3×3 matrix.

Likewise

$$\mathbf{GT} + \mathbf{JZ} = [g \ h]\begin{bmatrix}t\\w\end{bmatrix} + [j][z],$$

$$= [gt + hw + jz].$$

which is the bottom right hand term in the original product. Similarly the other terms work out correctly.

In other words, multiplication in blocks produces the proper answer. A fully rigorous proof of this depends upon what one demands. A formal proof by manipulation of suffixes is very tedious, and the type of verification we have just given will suffice for our purposes.

Blocks are very useful when finding powers of stochastic matrices. With these the problem is always to find powers of matrices of the form

$$\mathbf{P} = \begin{bmatrix}\mathbf{Q} & \mathbf{R}\\ \mathbf{O} & \mathbf{I}\end{bmatrix}.$$

Experimenting,

$$\mathbf{P}^2 = \begin{bmatrix}\mathbf{Q}^2 & \mathbf{QR} + \mathbf{R}\\ \mathbf{O} & \mathbf{I}\end{bmatrix}, \quad \mathbf{P}^3 = \begin{bmatrix}\mathbf{Q}^3 & (\mathbf{Q}^2 + \mathbf{Q} + \mathbf{I})\mathbf{R}\\ \mathbf{O} & \mathbf{I}\end{bmatrix},$$

and so it is plausible that

$$\mathbf{P}^n = \begin{bmatrix}\mathbf{Q}^n & (\mathbf{I} + \mathbf{Q} + \mathbf{Q}^2 + \ldots + \mathbf{Q}^{n-1})\mathbf{R}\\ \mathbf{O} & \mathbf{I}\end{bmatrix}.$$

It is quite simple to prove this by induction.

In the example considered above

$$\mathbf{Q} = \begin{bmatrix}\frac{1}{3} & \frac{1}{2}\\ \frac{1}{2} & \frac{1}{3}\end{bmatrix}$$

and it is easily verified that $\mathbf{Q}^n \to \mathbf{O}$ as $n \to \infty$. It may be shown that this happens in other cases also.

By direct multiplication we see that the identity

$$(\mathbf{I} + \mathbf{Q} + \mathbf{Q}^2 + \ldots + \mathbf{Q}^{n-1})(\mathbf{I} - \mathbf{Q}) = \mathbf{I} - \mathbf{Q}^n$$

holds for matrices, just as in ordinary algebra. Multiplying each side of the equation on the right by $(\mathbf{I} - \mathbf{Q})^{-1}$, we have

$$\mathbf{I} + \mathbf{Q} + \mathbf{Q}^2 + \ldots + \mathbf{Q}^{n-1} = (\mathbf{I} - \mathbf{Q}^n)(\mathbf{I} - \mathbf{Q})^{-1}.$$

Hence, assuming that $\mathbf{Q}^n \to \mathbf{O}$,

$$\mathbf{P}^n \to \begin{bmatrix} \mathbf{O} & \mathbf{NR} \\ \mathbf{O} & \mathbf{I} \end{bmatrix}, \quad \text{where } \mathbf{N} = (\mathbf{I} - \mathbf{Q})^{-1}.$$

Applying this to Example 6.7

$$\mathbf{Q} = \begin{bmatrix} \frac{1}{3} & \frac{1}{2} \\ \frac{1}{2} & \frac{1}{3} \end{bmatrix}, \quad \mathbf{I} - \mathbf{Q} = \begin{bmatrix} \frac{2}{3} & \frac{1}{2} \\ \frac{1}{2} & \frac{2}{3} \end{bmatrix},$$

$$\mathbf{N} = (\mathbf{I} - \mathbf{Q})^{-1} = \frac{6}{7}\begin{bmatrix} 4 & 3 \\ 3 & 4 \end{bmatrix},$$

$$\mathbf{R} = \begin{bmatrix} \frac{1}{6} & 0 \\ 0 & \frac{1}{6} \end{bmatrix} \quad \text{and} \quad \mathbf{NR} = \begin{bmatrix} \frac{4}{7} & \frac{3}{7} \\ \frac{3}{7} & \frac{4}{7} \end{bmatrix}.$$

NR is the part of the matrix $\mathbf{P}^n$ which is concerned with the probabilities of making a transition from the various possible starting states to the various possible absorbing states. So it is easily seen that the odds are 4:3 in favour of the player who starts winning.

Exercise 6.4

A married couple decide to go on having children until they have a boy and then stop. Draw up a matrix of transition probabilities.

Exercise 6.5

I decide to keep betting on the red at roulette, starting always with a unit stake and doubling it if I lose. Draw up a matrix of transition probabilities. Consider the various assumptions which must be made if this is to be a good model of the real situation.

Exercise 6.6

The weather may be good, bad or indifferent. A rudimentary method of forecasting attributes varying probabilities to the state of the weather on a certain day, knowing the weather on the previous day. Draw up a matrix of plausible transition probabilities and experiment with it.

Exercise 6.7

Each night we have to order the milk for next morning. If we have a pint, or less, left we order two pints. If we have more than a pint left we order one pint. Clearly, if we ordered a pint yesterday it is more likely that we will have to order two pints today, than it is if we ordered two yesterday. We average ten pints per week. Draw up a plausible model of the situation by suggesting a possible matrix of the transition probabilities between the states 'one pint' or 'two pints'.

Exercise 6.8

A Cambridge examination paper contained the following example.
Three men play a gambling game in which a player tosses three pennies.

If there are three heads he wins outright.
If there are two heads he has another throw.
If there is one head he passes to the player on his right, who throws with the same rules.
If there are no heads he passes to the player on his left, who throws with the same rules.
What are the respective chances of the three players, when one particular player starts?

The next example shows how further interesting information may be derived from the matrix **N**. We will consider one particular problem, but solve it by a method which indicates a general result.

EXAMPLE 6.8

Consider a miniature game of Snakes and Ladders. For this game we have a die or a spinner on which you may throw a 1, 2 or 3 with equal probability. The game is played on a board numbered 1 to 6, with a ladder from 3 to 4 and a snake from 5 to 2 (see Fig. 6.3). We may there-

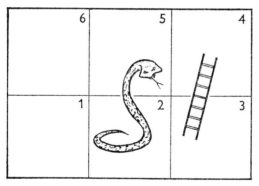

FIG. 6.3

fore analyse the game in terms of 5 states, 0 (meaning about to start), 1, 2, 4 and 6. State 6 is 'home'—an absorbing state. To give two examples of the game, throws of 3, 1, 2, 2 would take the player to states 4, 2, 4 and 6, whereas throws of 1, 3, 2 take him through states 1, 4 and 6.

The matrix of transition probabilities is

	0	1	2	4	6
0	0	$\frac{1}{3}$	$\frac{1}{3}$	$\frac{1}{3}$	0
1	0	0	$\frac{1}{3}$	$\frac{2}{3}$	0
2	0	0	$\frac{1}{3}$	$\frac{2}{3}$	0
4	0	0	$\frac{1}{3}$	0	$\frac{2}{3}$
6	0	0	0	0	1

We have seen how to calculate the probabilities of being in various states after various numbers of throws, but it is reasonable to ask the further question, 'How many throws will be needed, on the average, to finish the game?'

It will be convenient to use the notation $[\mathbf{P}]_{st}$ to denote the element in row s and column t of the matrix $\mathbf{P}$. Starting from state 0, consider a large number of trial games. We will assume that there are t trials, over which the various experimental possibilities average out.

This means that after 1 throw we are in state 1 on $\frac{1}{3}t$ occasions, state 2 on $\frac{1}{3}t$ occasions and state 4 on $\frac{1}{3}t$ occasions. More generally, after 1 throw we are in state k on $t[\mathbf{P}]_{0k}$ occasions. We may limit attention for the moment to $k = 0, 1, 2, 4$, which means that we need only consider the sub-matrix $\mathbf{Q}$, and we may say:

After 1 throw we are in state k on $t[\mathbf{Q}]_{0k}$ occasions;
Likewise, after 2 throws we are in state k on $t[\mathbf{Q}^2]_{0k}$ occasions;
Also, after 3 throws we are in state k on $t[\mathbf{Q}^3]_{0k}$ occasions; etc.

Hence the total number of throws which terminate in state k is

$$t[\mathbf{Q} + \mathbf{Q}^2 + \mathbf{Q}^3 + \ldots]_{0k}.$$

It follows that the average number of throws which terminate in state k, per game is

$$[\mathbf{Q} + \mathbf{Q}^2 + \mathbf{Q}^3 + \ldots]_{0k}.$$

To obtain the average number of throws needed to finish the game we must add these terms for $k = 1, 2, 4$, and also add in the single throw which takes us to the absorbing state 6. The addition of the one can be handled by inserting an additional term, the unit matrix, at the beginning of the bracket. In more general games some time might be spent waiting in state 0, but in this game no time is so spent and this is looked after by the zero terms in the first column of $\mathbf{P}$ and its sub-matrix $\mathbf{Q}$. Hence, starting in state 0, the average number of throws needed to get home in this case and in the more general case is given by the first row of the matrix product

$$[\mathbf{I} + \mathbf{Q} + \mathbf{Q}^2 + \ldots] \begin{bmatrix} 1 \\ 1 \\ 1 \\ \vdots \end{bmatrix}.$$

The other positions in this matrix product, which is of course a column vector, give the number of moves needed if the player starts in the other states 1, 2, or 4.

This product can be simplified since

$$\mathbf{I} + \mathbf{Q} + \mathbf{Q}^2 + \ldots = \mathbf{N}.$$

So using **e** to denote a column vector of ones we are led to the following result:

Ne is a column vector whose elements give the average number of throws before the process is absorbed, after starting in a given state. In this particular example the reader may verify that

$$\mathbf{N} = (\mathbf{I} - \mathbf{Q})^{-1} = \begin{bmatrix} 1 & \frac{1}{3} & \frac{17}{12} & \frac{3}{2} \\ 0 & 1 & \frac{5}{4} & \frac{3}{2} \\ 0 & 0 & \frac{9}{4} & \frac{3}{2} \\ 0 & 0 & \frac{3}{4} & \frac{3}{2} \end{bmatrix},$$

and

$$\mathbf{Ne} = (4{\cdot}25,\ 3{\cdot}75,\ 3{\cdot}75,\ 2{\cdot}25)'.$$

In particular, starting in state 0 the average number of throws needed to finish the game is 4·25.

EXAMPLE 6.9

A petrol company recently conducted an advertising campaign in which a plastic reproduction of the head of a footballer was given away with each purchase. There were 16 heads in the set. How many purchases had to be made on the average in order to collect a complete set? We must assume of course that the various footballers were randomly distributed in the petrol stations, and we can take no account of people making the task easier by swapping.

There are 17 states, which may be labelled 0 to 16, according to the number of footballers you already have. If you have r footballers already the probability that you will get a duplicate next time and remain in the same state is $r/16$, whereas the probability that you will get a new one and move into the next state is $(16 - r)/16$. The matrix of transition probabilities is therefore the 17×17 matrix

$$\begin{bmatrix} 0 & 16 & 0 & 0 & . & . & . \\ 0 & 1 & 15 & 0 & . & . & . \\ 0 & 0 & 2 & 14 & . & . & . \\ 0 & 0 & 0 & 3 & . & . & . \\ . & . & . & . & . & . & . \\ . & . & . & . & . & 15 & 1 \\ . & . & . & . & . & 0 & 1 \end{bmatrix} \div 16$$

The particularly simple form of matrix suggests that there may be other simple non-matrix methods of solving the problem (try to find one); it also suggests that inverting $\mathbf{I} - \mathbf{Q}$ may be especially easy this time. Remembering that **Q** is the 16×16 matrix in the top left-hand corner, it is easily seen that

$$\mathbf{N} = (\mathbf{I} - \mathbf{Q})^{-1} = 16 \begin{bmatrix} \frac{1}{16} & \frac{1}{15} & \frac{1}{14} & . & . & \frac{1}{2} & 1 \\ 0 & \frac{1}{15} & \frac{1}{14} & . & . & \frac{1}{2} & 1 \\ 0 & 0 & \frac{1}{14} & . & . & \frac{1}{2} & 1 \\ . & . & . & . & . & . & . \\ 0 & 0 & 0 & . & . & 0 & 1 \end{bmatrix}.$$

The average number of purchases required to go from state 0 to state 16 is therefore

$$16(1 + \tfrac{1}{2} + \tfrac{1}{3} + \ldots + \tfrac{1}{16}).$$

A good approximation to the term in the bracket can be obtained from the formula for Euler's limit,

$$1 + \tfrac{1}{2} + \tfrac{1}{3} + \ldots + \frac{1}{n-1} - \log n \to 0 \cdot 577 \ldots .$$

The bracket is therefore approximately

$$\log 17 + 0 \cdot 577 = 3 \cdot 41$$

and the average number of purchases required is about 54.

Barnard [2] describes an interesting use of stochastic matrices in the storage of confidential information. Students have recently seen the storage of information about individuals in a data bank as a threat to freedom. Barnard points out that the overall planning purposes for which the data is required are adequately met if the data is recorded with a known frequency of errors superimposed. For example, a student who achieves a third-class degree can be recorded as having a first, a second or a third, deciding which to record in some random way, with known probabilities which could be (say) 0·3, 0·2 and 0·5 respectively. A similar strategy, with different probabilities, would be used with students who had second-class or first-class degrees. These probabilities are entries in a stochastic matrix which transforms the true data into the fictitious data which is recorded. Subsequently if the aggregated data is transformed by the inverse matrix the statistical information which is required is obtained; the details on any particular individual have been irretrievably lost.

There is more to it than this, because the objection might be raised that it is far easier to aggregate results at the start and keep no record of individuals at all. But data of this kind is often used to calculate correlations—what proportion of people with first-class degrees went to highly paid jobs, for example, and the point of keeping the stochastically scrambled details is that, with suitable techniques, correlations which would be lost by a simple process of totalling can still be calculated.

Barnard also describes an ingenious method employed by a medical statistician in the USA to collect information about abortions. These are illegal in most of the USA and statisticians who learn that individual women have had abortions can be charged as accessories unless they communicate any knowledge they have of these individual cases to the District Attorney. We leave it to the reader to see if he can suggest a way of using a stochastic transformation which will enable the statistician to get his information whilst keeping within the law.

The next example shows an application of matrices to economic problems. An over-simplified example is considered in order that the arithmetic may be done 'by hand'. References are given later to some more serious examples.

EXAMPLE 6.10

Three families live on a small island. They are cut off from the outside world. The fathers of the families A, B and C, are respectively a farmer, a carpenter and a tailor. They live a happy life, sharing out the products of their labours. These, however, are not shared equally. In fact an analysis shows that the consumption of their products can be tabulated as follows.

	A	B	C
A	$\frac{1}{3}$	$\frac{1}{4}$	$\frac{1}{2}$
B	$\frac{1}{3}$	$\frac{1}{2}$	$\frac{3}{10}$
C	$\frac{1}{3}$	$\frac{1}{4}$	$\frac{1}{5}$

The meaning of the table is that farmer A's products are shared equally, whereas carpenter B works a quarter of the time for each of the other families and half the time for himself, and tailor C's products are given to the three families in proportions $\frac{1}{2}$, $\frac{3}{10}$, $\frac{1}{5}$. This table may be regarded as a stochastic matrix, but it is stochastic by columns whereas the earlier examples were stochastic by rows.

One day the islanders decide that they wish to be like any other civilized community and introduce money. You are asked to advise on a wages and incomes policy!

This is a good discussion example, and it is instructive not to run straight away into applying some mathematical technique, but to consider various economic possibilities. (For example, is it possible to run this system by decreeing that everybody shall have an equal income?) Clearly by enforcing sufficient arbitrary controls any system can be made to work. But if we decide that there is to be a free circulation of money, with 'fair' prices and no accumulation of capital (there being nothing else to spend it on) and no accumulation of debt then we must argue as follows.

Figure 6.4 shows schematically the flow of goods. Observe that the

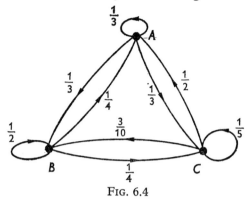

FIG. 6.4

numbers attached to the arrows are not in any common scale of measurement, they show merely how the various fractions of the output emanating from the three sources are disposed. Let us now attach *values x, y* and *z* to the goods which the three men produce. This leads to Fig. 6.5,

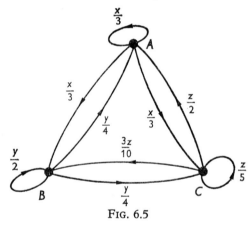

FIG. 6.5

and now it is reasonable to ensure that the inflows and outflows at each vertex balance. Balance can now be arranged because everything is measured in terms of a common monetary unit. There will be a flow of money along each arrow, in the opposite direction, to pay for the goods. Each individual will also have a certain capital represented by the money which he pays himself in order to balance the books.

For A's account to balance:
$$x = \tfrac{1}{3}x + \tfrac{1}{4}y + \tfrac{1}{2}z.$$

For B's account to balance:
$$y = \tfrac{1}{3}x + \tfrac{1}{2}y + \tfrac{3}{10}z;$$

and for C's to balance:
$$z = \tfrac{1}{3}x + \tfrac{1}{4}y + \tfrac{1}{5}z.$$
From this we find
$$x:y:z = 39:44:30.$$

The wealth of the three men is in the ratio $39:44:30$. If the matrix describing the transactions is called $\mathbf{M}$ it is easy to see that the vector which we have calculated is a column eigenvector $\mathbf{x}$, with eigenvalue unity. It satisfies the equation
$$\mathbf{Mx} = \mathbf{x}.$$

The above example shows in as simple a form as possible how matrices may be applied to economic problems. Models of this type are often called *Leontief models* after their inventor. A popular account of these methods may be found in an article by Stone [18]. This gives a 40×40 matrix description of the British economy in 1960, which elsewhere has been expanded to a 253×253 matrix model. Details of matrix calculations on the British economy may also be found in a government Blue Book [4]. Stone's more technical writings are available in [19].

In all of these problems the successive stages of evolution of the system are described by vectors, and each vector is derived from its predecessor by multiplying it by the stochastic matrix. Such a sequence of vectors is often called a *Markov chain* after the first mathematician to investigate them.

This concludes the section on stochastic matrices. A wide variety of examples of varying degrees of difficulty may be found in References [5, 6, 7, 11, 17].

6.6 Theory

Eigenvalues and eigenvectors were first introduced in Section 2.5. There we saw that if λ is an eigenvalue of a square matrix $\mathbf{A}$ and $\mathbf{x}$ a corresponding eigenvector (column eigenvector) they satisfy the relation
$$\mathbf{Ax} = \lambda\mathbf{x}.$$

The eigenvalues can be obtained as the roots of the characteristic equation
$$|\mathbf{A} - \lambda\mathbf{I}| = 0,$$
and the eigenvectors corresponding to the various eigenvalues can be found by solving the equations
$$(\mathbf{A} - \lambda\mathbf{I})\mathbf{x} = 0,$$
after substituting the appropriate values for λ.

A completely similar course of action can be carried out with row vectors instead of column vectors, and as an example we will find the row eigenvectors of the matrix

$$\mathbf{A} = \begin{bmatrix} -1 & 2 & 2 \\ -8 & 7 & 4 \\ -13 & 5 & 8 \end{bmatrix}$$

used in Section 2.5. A row eigenvector is a non-zero vector **y** which satisfies the relation

$$\mathbf{yA} = \lambda \mathbf{y}.$$

It is not immediately obvious that the eigenvalues associated with row eigenvectors must have the same numerical values as those associated with column eigenvectors, but the methods of Section 2.5 apply again and lead to exactly the same characteristic equation as before. Thus, if $\mathbf{y} = (y_1, y_2, y_3)$, the row eigenvector satisfies

$$[y_1, \; y_2, \; y_3] \begin{bmatrix} -1 & 2 & 2 \\ -8 & 7 & 4 \\ -13 & 5 & 8 \end{bmatrix} = \lambda [y_1, \; y_2, \; y_3].$$

Hence

$$(-y_1 - 8y_2 - 13y_3, \; 2y_1 + 7y_2 + 5y_3, \; 2y_1 + 4y_2 + 8y_3) = (\lambda y_1, \; \lambda y_2, \; \lambda y_3).$$

Writing this vector equation as three scalar equations

$$\begin{aligned} -y_1 - 8y_2 - 13y_3 &= \lambda y_1, \\ 2y_1 + 7y_2 + 5y_3 &= \lambda y_2, \\ 2y_1 + 4y_2 + 8y_3 &= \lambda y_3. \end{aligned}$$

Rearranging

$$\begin{aligned} (-1-\lambda)y_1 - 8y_2 - 13y_3 &= 0, \\ 2y_1 + (7-\lambda)y_2 + 5y_3 &= 0, \\ 2y_1 + 4y_2 + (8-\lambda)y_3 &= 0. \end{aligned} \tag{6.7}$$

Assuming once again the theorem that a set of homogeneous linear equations has a solution in which the unknowns are not all zero if and only if the determinant of the coefficients is zero we have

$$\begin{vmatrix} -1-\lambda & -8 & -13 \\ 2 & 7-\lambda & 5 \\ 2 & 4 & 8-\lambda \end{vmatrix} = 0. \tag{6.8}$$

This determinant is the transpose of the one in Eq. (2.21) of Section 2.5. Determinants (unlike matrices) are unaltered by transposition, so the expansion of Eq. (6.8) is exactly the same as the expansion of Eq. (2.21) which is, after a trivial change of sign,

$$\lambda^3 - 14\lambda^2 + 63\lambda - 90 = 0. \tag{6.9}$$

As before, the roots of this are
$$\lambda = 3, 5 \text{ or } 6.$$

When $\lambda = 3$ Eqs. (6.7) become

$$\begin{aligned} -4y_1 - 8y_2 - 13y_3 &= 0, \\ 2y_1 + 4y_2 + 5y_3 &= 0, \\ 2y_1 + 4y_2 + 5y_3 &= 0, \end{aligned} \tag{6.10}$$

which have solutions

$$y_1 : y_2 : y_3 = -2 : 1 : 0.$$

This gives the row eigenvector with eigenvalue 3, and we may put

$$\mathbf{y}_1 = (-2, 1, 0).$$

Note carefully how the equations (6.10) arise. It is much more convenient to print a set of equations by listing them down the page rather than spreading them across, but this layout might at first sight suggest that the matrix **A** has been transposed. This, however, is not a good way of looking at it, although it would be correct to say that the row eigenvectors of **A** may be obtained by calculating the column eigenvectors of **A**′.

Putting $\lambda = 5$ in Eqs. (6.7) we obtain eventually

$$\mathbf{y}_2 = (7, -2, -2).$$

Putting $\lambda = 6$,

$$\mathbf{y}_3 = (-3, 1, 1).$$

The column eigenvectors which were calculated in Section 2.5 may be written

$$\begin{aligned} \mathbf{x}_1 &= (0, 1, -1)' & \text{with eigenvalue } \lambda = 3, \\ \mathbf{x}_2 &= (1, 2, 1)' & \text{with eigenvalue } \lambda = 5, \end{aligned}$$

and

$$\mathbf{x}_3 = (2, 4, 3)' \quad \text{with eigenvalue } \lambda = 6.$$

In Theorem 2.3 it was shown that column vectors associated with different eigenvalues of a symmetric matrix are orthogonal. A short calculation easily verifies that this is not the case with this non-symmetric matrix. The orthogonality is, however, maintained in a modified form. Experiment shows that

$$\mathbf{y}_1 \mathbf{x}_2 = [-2, 1, 0] \begin{bmatrix} 1 \\ 2 \\ 1 \end{bmatrix} = 0,$$

and likewise

$$\mathbf{y}_r \mathbf{x}_s = 0 \quad \text{whenever } r \neq s.$$

It is easy to prove in general that given any matrix $\mathbf{A}$, any column eigenvector corresponding to a particular eigenvalue is orthogonal to all row eigenvectors corresponding to other eigenvalues; and vice versa. The proof is as follows.

Let λ and μ be distinct eigenvalues, and let $\mathbf{x}$ be a column eigenvector with eigenvalue λ, and let $\mathbf{y}$ be a row eigenvector with eigenvalue μ. Then

$$\mathbf{A}\mathbf{x} = \lambda \mathbf{x} \quad \text{and} \quad \mathbf{y}\mathbf{A} = \mu \mathbf{y}.$$

Hence

$$\mathbf{y}\mathbf{A}\mathbf{x} = \lambda \mathbf{y}\mathbf{x} \quad \text{and} \quad \mathbf{y}\mathbf{A}\mathbf{x} = \mu \mathbf{y}\mathbf{x}.$$

Hence

$$\lambda \mathbf{y}\mathbf{x} = \mu \mathbf{y}\mathbf{x}$$

and if $\lambda \neq \mu$

$$\mathbf{y}\mathbf{x} = 0.$$

In this example the vectors have the further property

$$\mathbf{y}_r \mathbf{x}_r = 1 \quad \text{for } r = 1, 2, 3.$$

This has arisen because the example was constructed so that the numbers worked out easily, but this can always be arranged because eigenvectors are determined only as far as the ratios of their components, and they may be scaled to suit any other purpose which we have in mind. In Section 5.1 the eigenvectors were normalized, that is scaled so that $\mathbf{x}'\mathbf{x} = 1$. In the present context we may scale the row eigenvector or the column eigenvector (or both) so that $\mathbf{y}\mathbf{x} = 1$.

We may now adapt the work of Section 5.1 to asymmetric matrices, by forming the matrices which were then called $\mathbf{A}_1$, $\mathbf{A}_2$ and $\mathbf{A}_3$. Accordingly we now define

$$\mathbf{A}_1 = \mathbf{x}_1 \mathbf{y}_1 = \begin{bmatrix} 0 \\ 1 \\ -1 \end{bmatrix} \begin{bmatrix} -2 & 1 & 0 \end{bmatrix} = \begin{bmatrix} 0 & 0 & 0 \\ -2 & 1 & 0 \\ 2 & -1 & 0 \end{bmatrix},$$

$$\mathbf{A}_2 = \mathbf{x}_2 \mathbf{y}_2 = \begin{bmatrix} 1 \\ 2 \\ 1 \end{bmatrix} \begin{bmatrix} 7 & -2 & -2 \end{bmatrix} = \begin{bmatrix} 7 & -2 & -2 \\ 14 & -4 & -4 \\ 7 & -2 & -2 \end{bmatrix},$$

and

$$\mathbf{A}_3 = \mathbf{x}_3 \mathbf{y}_3 = \begin{bmatrix} 2 \\ 4 \\ 3 \end{bmatrix} \begin{bmatrix} -3 & 1 & 1 \end{bmatrix} = \begin{bmatrix} -6 & 2 & 2 \\ -12 & 4 & 4 \\ -9 & 3 & 3 \end{bmatrix}$$

The following relations now hold as in Exercise 5.2,

$$\mathbf{A}_r\mathbf{A}_s = 0, \quad r \neq s,$$
$$\mathbf{A}_r\mathbf{A}_r = \mathbf{A}_r, \quad r = 1, 2, 3.$$

$$\mathbf{A}_1 + \mathbf{A}_2 + \mathbf{A}_3 = \mathbf{I}, \tag{6.11}$$

$$\lambda_1\mathbf{A}_1 + \lambda_2\mathbf{A}_2 + \lambda_3\mathbf{A}_3 = \mathbf{A}. \tag{6.12}$$

Exercise 6.9

Verify these relations in this particular case, and demonstrate also that matrices $\mathbf{A}_1, \mathbf{A}_2, \mathbf{A}_3, \ldots$ may be constructed with these properties, starting with any matrix $\mathbf{A}$ with distinct eigenvalues.

When a matrix has repeated eigenvalues it may or may not possess a spectral decomposition of this kind. It can be shown that a necessary and sufficient condition for this to be possible is for $\mathbf{A}$ to be *normal*, that is to possess the property $\mathbf{A}'\mathbf{A} = \mathbf{A}\mathbf{A}'$. Normal matrices are explained in Gantmacher [10].

The above equations are frequently useful, and we can use them now to demonstrate what happens when successive powers of a matrix are taken.

Raising both sides of Eq. (6.12) to the nth power, and using the orthogonal relations

$$\lambda_1^n\mathbf{A}_1 + \lambda_2^n\mathbf{A}_2 + \lambda_3^n\mathbf{A}_3 = \mathbf{A}^n.$$

If one eigenvalue has a larger numerical value than the others then, for large values of n, the term on the left-hand side containing it dominates; hence if λ_3 is the root of largest modulus, as $n \to \infty$

$$\mathbf{A}^n \sim \lambda_3^n\mathbf{A}_3.$$

(The beginner may take the $\sim$ sign to mean 'approximates to'; the expert can give it a more precise definition.)

Furthermore if $\mathbf{x}$ is an arbitrary column vector

$$\mathbf{A}^n \mathbf{x} \sim \lambda_3^n \mathbf{A}_3 \mathbf{x}, \quad n \to \infty.$$

Remembering that $\mathbf{A}_3 = \mathbf{x}_3\mathbf{y}_3$,

$$\mathbf{A}^n\mathbf{x} \sim \lambda_3^n\mathbf{x}_3\mathbf{y}_3\mathbf{x}, \quad n \to \infty;$$

and $\mathbf{y}_3\mathbf{x}$ (the product of row and column vectors) is a scalar constant, the exact value of which depends on the arbitrary vector $\mathbf{x}$ selected. Hence

$$\mathbf{A}^n\mathbf{x} \sim \lambda^n\mathbf{x}_3 \times \text{constant}, \quad n \to \infty.$$

Likewise

$$\mathbf{y}\mathbf{A}^n \sim \lambda_3^n\mathbf{y}_3 \times \text{constant}, \quad n \to \infty.$$

This explains many things, among them the phenomena occurring in the examples discussed at the beginning of this chapter. It also provides a feasible way of computing eigenvalues and eigenvectors. Given a matrix, by iterating on an arbitrary vector we approach the eigenvector associated with the dominant eigenvalue (i.e. the eigenvalue of largest modulus). This may be done with column vectors or row vectors. Also, for large n,

$$\frac{\mathbf{A}^{n+1}\mathbf{x}}{\mathbf{A}^n\mathbf{x}} \sim \lambda_{\max}.$$

This enables the eigenvalue of largest absolute value to be calculated without forming and solving the characteristic equation. It must not be thought that this method is 'approximate' whereas the characteristic equation method is 'exact', because in practice polynomial equations can only very rarely be solved by methods which are exact, far more often than not they have to be solved by a method which gives an approximation to a root to a known degree of accuracy.

Having found one eigenvalue and the associated eigenvector there are various ways in which others may be found as well. Continuing to write in terms of the 3×3 example, if λ_3 and the related row and column eigenvectors have been calculated this enables $\mathbf{A}_3$ to be calculated, and we may then form the matrix

$$\mathbf{B} = \mathbf{A} - \lambda_3\mathbf{A}_3 = \lambda_1\mathbf{A}_1 + \lambda_2\mathbf{A}_2.$$

It is easy to confirm that $\mathbf{B}$ has eigenvalues 0, λ_1, λ_2 and the same eigenvectors as $\mathbf{A}$. Iteration with $\mathbf{B}$ then enables λ_2, $\mathbf{x}_2$ and $\mathbf{y}_2$ to be found, where λ_2 is the eigenvalue of next largest modulus.

Continuing in this way, in principle all the eigenvalues of a matrix can be found if they are distinct; however, there are practical difficulties with this method. Repeated eigenvalues produce further complications which we cannot discuss here, but which are important in practice.

Another possible method arises as follows. If λ is any number, using Eq. (6.11) and (6.12),

$$(\lambda_1 - \lambda)\mathbf{A}_1 + (\lambda_2 - \lambda)\mathbf{A}_2 + (\lambda_3 - \lambda)\mathbf{A}_3 = \mathbf{A} - \lambda\mathbf{I}.$$

Using the orthogonal relations it may be verified that

$$(\lambda_1 - \lambda)^{-1}\mathbf{A}_1 + (\lambda_2 - \lambda)^{-1}\mathbf{A}_2 + (\lambda_3 - \lambda)^{-1}\mathbf{A}_3 = (\mathbf{A} - \lambda\mathbf{I})^{-1}.$$

The left-hand side of this equation is the spectral decomposition of the matrix $(\mathbf{A} - \lambda\mathbf{I})^{-1}$, which has eigenvalues $(\lambda_1 - \lambda)^{-1}$, $(\lambda_2 - \lambda)^{-1}$ and $(\lambda_3 - \lambda)^{-1}$. By suitable choice of λ any one of these can be made dominant, and so if λ is near to λ_2 (say) iteration with $(\mathbf{A} - \lambda\mathbf{I})^{-1}$ will enable $(\lambda_2 - \lambda)^{-1}$ to be calculated and hence λ_2. (Compare resonance in mechanics!)

It was stated earlier that when a given matrix is used to multiply an arbitrary vector repeatedly the resulting sequence of vectors tends to the eigenvector associated with the eigenvalue of largest modulus. This is so unless the arbitrary vector happens by chance to be orthogonal to the dominant eigenvector, in which case the sequence tends to the eigenvector associated with the next eigenvalue in order of magnitude.

However, in numerical computation this is unlikely to happen even if the initial vector is (theoretically) orthogonal to the dominant eigenvector, because rounding errors in the subsequent calculation will be sufficient to 'tip it off balance'.

This brief and incomplete discussion should be sufficient to indicate something of the methods which are available for the numerical computation of eigenvalues and eigenvectors. This is a very important aspect of numerical mathematics which has been developed extensively in recent years. Notice that these methods do not depend on the use of determinants at all.

Exercise 6.10
Show that the matrix

$$\mathbf{A} = \begin{bmatrix} 0 & \frac{1}{2} & 0 \\ 1 & 0 & \frac{1}{3} \\ 3 & 0 & 0 \end{bmatrix}$$

which occurs in the second beetle problem in Example 6.2, has eigenvalues 1, $\frac{1}{2}(-1 \pm i)$. Denoting $\frac{1}{2}(-1 \pm i)$ by α and β show that the corresponding row eigenvectors are $(6, 3, 1)$, $(6\alpha^2, 3\alpha, 1)$, $(6\beta^2, 3\beta, 1)$ and that the column eigenvectors are $(1, 2, 3)'$, $(\alpha, 2\alpha^2, 3)'$, $(\beta, 2\beta^2, 3)'$. Obtain the spectral decomposition of $\mathbf{A}$ and verify that starting, as in the earlier example, with an initial population of $(1000, 1000, 1000)$ the equilibrium population is $(2400, 1200, 400)$.

Exercise 6.11
Show that the matrix

$$\mathbf{A} = \begin{bmatrix} 0 & 1 & 1 \\ 0 & 0 & 1 \\ 1 & 1 & 0 \end{bmatrix}$$

which describes the chemical reaction in Example 6.3, has eigenvalues γ, -1, $1-\gamma$, where γ denotes the golden ratio number 1·618 (approximately), which is the positive root of the equation $x^2 = x + 1$. Show that γ is the dominant eigenvalue, and compute the related row and column eigenvectors. In what ratios are the O, OH and H radicals produced in the chemical reaction in the long run?

Exercise 6.12
Consider the interpretation of the multiplication of a vector by a symmetric matrix given in Fig. 2.8 of Section 2.7. When the multiplying matrix is symmetric interpret the recent results on iteration from a geometrical point of view.

It has been convenient to display the spectral decomposition of a matrix and to state its related properties in a notation which brings out the connexions with Fourier series and eigenfunctions in vector spaces of infinite dimension, discussed in Chapter 5. However, in other contexts there is no advantage in doing this, and we next suggest that the reader may investigate writing the results in an alternative notation. This will also provide useful experience in handling block matrices.

Exercises 6.13

Consider a 3×3 matrix $\mathbf{A}$ with distinct eigenvalues $\lambda_1, \lambda_2, \lambda_3$. Let the related column eigenvectors be $\mathbf{x}_1, \mathbf{x}_2, \mathbf{x}_3$, and the related row eigenvectors be $\mathbf{y}_1, \mathbf{y}_2, \mathbf{y}_3$. Normalize so that

$$\mathbf{y}_r \mathbf{x}_r = 1; \quad r = 1, 2, 3.$$

Assemble the row eigenvectors into a matrix $\mathbf{Y}$,

$$\mathbf{Y} = \begin{bmatrix} \mathbf{y}_1 \\ \hdashline \mathbf{y}_2 \\ \hdashline \mathbf{y}_3 \end{bmatrix},$$

and the column eigenvectors into a matrix $\mathbf{X}$, $\mathbf{X} = [\mathbf{x}_1 \mid \mathbf{x}_2 \mid \mathbf{x}_3]$. Verify that $\mathbf{YX} = \mathbf{I}$. From known properties of inverse matrices it follows that $\mathbf{XY} = \mathbf{I}$ also. Verify that

$$\mathbf{YAX} = \operatorname{diag}\{\lambda_1, \lambda_2, \lambda_3\},$$

where the matrix on the right-hand side is a matrix with $\lambda_1, \lambda_2, \lambda_3$ on the principal diagonal and zeros elsewhere. Show that

$$\mathbf{YA}^n\mathbf{X} = \operatorname{diag}\{\lambda_1^n, \lambda_2^n, \lambda_3^n\},$$

that

$$\mathbf{A}^n = \mathbf{X} \operatorname{diag}\{\lambda_1^n, \lambda_2^n, \lambda_3^n\} \mathbf{Y},$$

and determine the behaviour of $\mathbf{A}^n$ and $\mathbf{A}^n\mathbf{x}$ for large n, where $\mathbf{x}$ is an arbitrary column vector.

6.7 Root processes continued

The square root process at the beginning of this chapter may be regarded as a particular case of a more general method, which will be introduced by means of examples and then discussed from a more theoretical point of view. This method, due to McDougall [5], is an iterative one which is easily carried out on machines, although it appears that it does not usually converge very quickly.

To find the cube root of 2 we may draw up a matrix

$$\begin{bmatrix} y & 2 & 2x \\ x & y & 2 \\ 1 & x & y \end{bmatrix}.$$

It can be seen that the matrix is constructed by building up diagonals. In the bottom left-hand corner a one is entered, and this will always be done. Along the diagonal above x's are entered, where x is a guessed approximation to $\sqrt[3]{2}$. Along the next diagonal, which happens in this case to be the principal diagonal, y's are entered, where y is a guessed approximation to $\sqrt[3]{2^2}$. Along the next diagonal 2's are entered and along the next, which has space only for one term, $2x$ is entered. We may take $x = 1$ and $y = 2$, getting

$$\mathbf{A} = \begin{bmatrix} 2 & 2 & 2 \\ 1 & 2 & 2 \\ 1 & 1 & 2 \end{bmatrix},$$

and we may either take powers of this matrix or iterate with it, repeatedly multiplying an arbitrary column vector. It will be found that the terms of any column of $\mathbf{A}^n$, or alternatively the terms of the arbitrary column vector, tend to the ratio

$$2^{2/3} : 2^{1/3} : 1.$$

Hence we have a method for calculating $2^{1/3}$ iteratively.

If $\sqrt[4]{5}$ is required the following matrix may be used

$$\begin{bmatrix} 4 & 5 & 5 & 10 \\ 2 & 4 & 5 & 5 \\ 1 & 2 & 4 & 5 \\ 1 & 1 & 2 & 4 \end{bmatrix}.$$

Here there is no choice about the one in the bottom left-hand corner. The 1's in the diagonal above are an approximation to $5^{1/4}$, and the 2's above this are an approximation to $5^{1/2}$, above this are the 4's which are an approximation to $5^{3/4}$, and the next diagonal consists of 5's exactly. The next diagonal contains 5's as an approximation to $5^{5/4}$, and these are 'forced' as it were by having chosen 1 as the approximate value of $5^{1/4}$; if 2 had been chosen this diagonal would have to contain 10's. The final diagonal, of one term only, contains 10, this being obtained by multiplying the previous approximation to $5^{1/2}$ by 5.

Once again if powers are taken, or if the matrix is used to multiply repeatedly an arbitrary column vector, the ratios of the terms in the columns tend to $5^{3/4} : 5^{1/2} : 5^{1/4} : 1$. (These calculations may be performed quite quickly with a desk calculator.)

It is clear that some explanation of this behaviour is required. This can be obtained by applying the results of the last section of the formulae in Section 4.4, where it was shown that the eigenvalues of the $n \times n$ circulant

$$\mathbf{C} = \begin{bmatrix} a & b & c & . & . & f & g \\ g & a & b & . & . & e & f \\ f & g & a & . & . & d & e \\ . & . & . & . & . & . & . \\ b & c & d & . & . & g & a \end{bmatrix}$$

are

$$\lambda = a + b\omega + c\omega^2 + \ldots + g\omega^{n-1}, \qquad \omega^n = 1,$$

the n nth roots of unity giving the n eigenvalues.

The matrices used in the root methods just discussed are not circulants, but circulant theory is easily extended to cover them. Thus adopting the arguments used previously on circulants we see, continuing with the particular example, that the matrix

$$\mathbf{A} = \begin{bmatrix} 4 & 5 & 5 & 10 \\ 2 & 4 & 5 & 5 \\ 1 & 2 & 4 & 5 \\ 1 & 1 & 2 & 4 \end{bmatrix}$$

has as a column eigenvector $(5^{3/4}, 5^{1/2}, 5^{1/4}, 1)'$ and this is why the method works. The associated eigenvalue is $5^{3/4} + 5^{1/2} + 2.5^{1/4} + 4$. (Check this.)

The other eigenvectors are of the form $(5^{3/4}\omega^3, 5^{1/2}\omega^2, 5^{1/4}\omega, 1)'$, with eigenvalues $5^{3/4}\omega^3 + 5^{1/2}\omega^2 + 2.5^{1/4}\omega + 4$, where $\omega^4 = 1$. (Check again.) By considering the appropriate diagram in the Argand plane it is easy to see that $5^{3/4} + 5^{1/2} + 2.5^{1/4} + 4$ is the eigenvalue of largest modulus, and so iteration with the matrix produces the corresponding eigenvector.

Here is an example of how matrix iteration may be used to solve more general algebraic equations. The eigenvalues of a matrix are the roots of the corresponding characteristic equation; but it may be easier to find the eigenvalues by some iterative technique with matrices than it is to solve the algebraic equation. Hence, given a polynomial equation it is reasonable to ask: 'Is it possible to obtain roots of the equation by finding a matrix which has the given polynomial as its characteristic polynomial, and iterating with the matrix?' This may be pursued as an investigation exercise.

Exercise 6.14

Verify that the matrix

$$\begin{bmatrix} a_1 & -1 & 0 & 0 \\ a_2 & 0 & -1 & 0 \\ a_3 & 0 & 0 & -1 \\ a_4 & 0 & 0 & 0 \end{bmatrix}$$

has the characteristic equation

$$\lambda^4 - a_1\lambda^3 + a_2\lambda^2 - a_3\lambda + a_4 = 0.$$

Consider various numerical values for a_1, a_2, a_3, a_4, and investigate the possibilities of obtaining a root of the equation by iterating with the matrix.

6.8 Problems involving continuous growth

An important idea of elementary analysis, frequently covered in school courses, is the idea of continuous growth regarded as the limiting case of discrete growth in geometrical progression. This is conveniently explained in terms of the growth of biological populations, in terms of the growth and decay of electrical charge (for example on the plate of a capacitor), or in terms of the growth of money at compound interest.

If £1 is invested at a rate of 5%, and the interest is calculated once per year only, then at the end of one year the £1 deposited has grown to £1·05. If interest is calculated every six months, however, it becomes

$$£\left(1 + \frac{0·05}{2}\right)$$

after six months, and this becomes

$$£\left(1 + \frac{0·05}{2}\right)^2 = £1·050625$$

after another six months. The depositor gains slightly as a result.

If interest is added every month, at the end of the year the depositor has

$$£\left(1 + \frac{0·05}{12}\right)^{12} = £1·051162.$$

If it is added every week, at the end of a year he has

$$£\left(1 + \frac{0·05}{52}\right)^{52} = £1·051246.$$

If interest is added continually then we need to evaluate the limit

$$\lim_{m \to \infty} \left(1 + \frac{0·05}{m}\right)^m.$$

Since

$$\left(1 + \frac{0·05}{m}\right)^m = \left(1 + \frac{1}{n}\right)^{n \times 0·05} \quad \text{if } m = n \times 0·05,$$

$$= \left[\left(1 + \frac{1}{n}\right)^n\right]^{0·05},$$

for theoretical purposes it is most convenient to calculate

$$\lim_{n \to \infty} \left(1 + \frac{1}{n}\right)^n$$

once and for all, and to derive the continuous growth rates corresponding to other rates of interest and periods of conversion from it. This limit is denoted by the letter e, and is well known as the base of natural logarithms and as a number with manifold mathematical applications.

Similar calculations may be based on the growth problems discussed earlier in Section 6.2. There a population (beetles, chemical compounds, etc.) was described by a vector, and over successive intervals of time the vectors grew in geometric progression—using this term in the extended sense to describe the repeated multiplication of the initial vector by a constant matrix. A limiting argument could be carried out for these cases starting from first principles, but it is more convenient to assume knowledge of the exponential function and apply it.

It is first necessary to formulate the general problem. We may consider a population divided into age groups in which birth and death is occurring, as in the beetle examples, or we may consider a population involving a number of types of animal which prey upon one another, or we may consider a reaction in which certain chemical substances change into others. In these, and other similar circumstances, it may be reasonable to construct a mathematical model in which we assume that the growth rate of any one component of the system depends linearly on the amounts of the other components present at the particular time. If this assumption is valid then whether we are considering animal populations, chemical reactions, economic exchanges of goods or the flow of electricity, the relevant equations are of the form

$$\begin{aligned}
\dot{x}_1 &= a_{11}x_1 + a_{12}x_2 + \ldots + a_{1n}x_n, \\
\dot{x}_2 &= a_{21}x_1 + a_{22}x_2 + \ldots + a_{2n}x_n, \\
&\;\;\vdots \\
\dot{x}_n &= a_{n1}x_1 + a_{n2}x_2 + \ldots + a_{nn}x_n.
\end{aligned}$$

x_r denotes the quantity of the rth component present, a dot denotes differentiation with respect to time and a_{rs} denotes the effect which the sth component has on the growth of the rth component. In matrix form these equations may be written

$$\dot{\mathbf{x}} = \mathbf{A}\mathbf{x}. \qquad (6.13)$$

We first seek solutions of this equation which have a particularly simple form (compare Section 3.1). Under what circumstances is the vector

$$\mathbf{x} = \mathbf{p}\, e^{\lambda t}, \qquad p_1, p_2, \ldots p_n \text{ constant} \qquad (6.14)$$

a solution? Substituting expression (6.14) in Eq. (6.13),

$$\lambda \mathbf{p}\, e^{\lambda t} = \mathbf{A}\mathbf{p}\, e^{\lambda t},$$

and so

$$\mathbf{A}\mathbf{p} = \lambda \mathbf{p}.$$

Once again we are led to study the eigenvalues and eigenvectors of the matrix **A**. Corresponding to each eigenvector we get a solution of the equation. It is easily verified that solutions may be added and the vectors

$$\mathbf{p}_1 e^{\lambda_1 t},\ \mathbf{p}_2 e^{\lambda_2 t},\ \ldots,\ \mathbf{p}_n e^{\lambda_n t}$$

form a basis for the vector space of solutions. As before $\mathbf{p}_r$ denotes the rth eigenvector of the matrix **A**, and not the rth component of vector **p**, which is denoted by p_r.

A number of theoretical questions remain. It is necessary to demonstrate that all of the solutions of Eq. (6.13) can be expressed as linear sums of these base vectors, and complications can arise when **A** has repeated eigenvalues. A detailed study of elementary examples at school level is given by Brand and Sherlock [5]. Chemical applications are included in the work by Noble [15] and economic applications in that by Allen [1].

Exercise 6.15

An alternative approach is to define the exponential function of the matrix **A** by the equation

$$e^{\mathbf{A}} = \mathbf{I} + \mathbf{A} + \frac{1}{2!}\mathbf{A}^2 + \frac{1}{3!}\mathbf{A}^3 + \ldots\ .$$

Verify that the solution of the equation $\dot{\mathbf{x}} = \mathbf{A}\mathbf{x}$ is given by

$$\mathbf{x} = e^{\mathbf{A}t}\mathbf{x}_0,$$

where $\mathbf{x}_0$ denotes the value of **x** at time $t = 0$. Verify also that the solution to the non-homogenous equation

$$\dot{\mathbf{x}} = \mathbf{A}\mathbf{x} + \mathbf{f}(t)$$

is given by

$$\mathbf{x} = e^{\mathbf{A}t}\mathbf{x}_0 + \int_0^t e^{\mathbf{A}(t-\tau)}\mathbf{f}(\tau)\, d\tau.$$

Equations of order higher than the first are just as easily handled as they may be reduced to special cases of the previous problem. Thus, given the equation

$$\frac{d^3x}{dt^3} + a_3 \frac{d^2x}{dt^2} + a_2 \frac{dx}{dt} + a_1 x = 0 \qquad (6.15)$$

we may put

$$x = x_1, \quad \frac{dx}{dt} = x_2 \quad \text{and} \quad \frac{d^2x}{dt^2} = x_3.$$

The latter equations with Eq. (6.15) give

$$\begin{aligned}\dot{x}_1 &= x_2, \\ \dot{x}_2 &= x_3, \\ \dot{x}_3 &= -a_1 x_1 - a_2 x_2 - a_3 x_3.\end{aligned} \quad (6.16)$$

The system (6.16) is equivalent to Eq. (6.15) and it can be handled by the previous methods.

References

1. ALLEN, R. G. D., *Mathematical Economics*, Macmillan (1959).
2. BARNARD, G. A., 'Mathematics and our Social Problems', *Bull. I.M.A.*, **7**, 1, pp. 7–10 (Jan. 1971).
3. BERNARDELLI, H., 'Population waves', *J. Burma Res. Soc.*, **31**, No. 1, 1–18.
4. BLUE BOOK, *National Income and Expenditure*, HMSO (1966).
5. BRAND, T. E. and SHERLOCK, A. J., *Matrices, Pure and Applied*, Arnold (1970).
6. DURRAN, J. H., *Statistics and Probability*, CUP (1970).
7. FELLER, W., *Introduction to Probability Theory and its applications*, Vol. 1, Wiley (1950).
8. FLETCHER, T. J., 'Combining matrices', *Mathl. Gaz.* LII, 379, 23–30 (Feb. 1968).
9. FLETCHER, T. J., 'A heuristic approach to matrices', *Educ. Studies in Maths.*, **1**, 166–80; Reidel, Holland (1968).
10. GANTMACHER, F. R., *The Theory of Matrices*, Vol. 1, Chelsea (1959).
11. KEMENY, SNELL and THOMPSON, *Introduction to Finite Mathematics*, Prentice Hall (1956).
12. LESLIE, P. H., 'On the use of matrices in certain population mathematics', *Biometrika*, XXXIII pt. III, 183–212 (1945).
13. LESLIE, P. H., 'Some further notes on the use of matrices in population mathematics', *Biometrika*, XXXV, pts. III and IV, 213–45 (1948).
14. MATTHEWS, G., *Matrices 2*, Arnold (1964).
15. NOBLE, B., *Applications of Undergraduate Mathematics in Engineering*, Macmillan, New York (1967).
16. REDFERN, P., *Input–output analysis and its applications to education and manpower planning*, C.A.S. occasional Paper 5, HMSO (1967).

17. SNELL, J. L., 'Finite Markov chains and their application', *Am. math. Mon.*, **66,** 2, 99–104 (1959).
18. STONE, R., 'Mathematics in the Social Sciences', *Scient. Am.* 168–82 (September 1964).
19. STONE, R., *Mathematics in the Social Sciences and Other Essays*, Chapman and Hall (1966).

7
Statistical Applications

There are many applications of linear algebra in statistics. This is not immediately apparent to the student, who often finds that the opening stages of statistics are concerned with means, standard deviations and tests of significance of various kinds. These highly important notions involve somewhat specialized mathematical considerations all of their own; but later on in statistics, in multivariate analysis, the situation changes, for the student who knows a little linear algebra finds much that has already become familiar in other contexts. This we will now try to show. Many details of great importance to the practising statistician will be ignored as our aim is merely to show how much of the linear algebra which has occurred earlier in the book can be put to new uses.

A constantly recurring theme is the emergence of matrices of the form $\mathbf{M'M}$. These matrices, which are sometimes called *covariance* or *correlation* matrices, according to circumstance, are the link between the various sections of this chapter.

7.1 Lines of best fit

It frequently occurs in science that a set of experimental readings is obtained, and there is reason to suppose that the readings when plotted should lie on a straight-line graph; but observations being liable to error they seldom do. How can one find the straight-line graph which fits the data best?

Suppose we have the three points (2, 3), (4, 5) and (7, 7), as shown in Fig. 7.1, and we seek the values of m and c which ensure that the line $y = mx + c$ fits the data as closely as possible. It is first necessary to decide what is meant by 'as closely as possible', because different interpretations can be put on these words. One possible interpretation is to try to arrange that the sum of the squares of the errors in the values of y, for given values of x, is as small as possible. Other criteria can be adopted, but the present one is often employed and it involves an interesting application of some of the ideas of linear algebra.

If the relation $y = mx + c$ is used to determine the values of y the three errors are $2m + c - 3$, $4m + c - 5$, and $7m + c - 7$ respectively. We wish to choose m and c so as to minimize

$$(2m + c - 3)^2 + (4m + c - 5)^2 + (7m + c - 7)^2$$
$$= 69m^2 + 26mc + 3c^2 - 150m - 30c + 83.$$

Denoting this expression by $F(m, c)$ a geometrical argument may now be employed. For any, d $F(m, c) = d$ is a curve of the second degree in the (m, c) plane. (The (m, c) plane is *not* the same plane as the (x, y) plane in which we started.) Since $4 \times 69 \times 3 > 26 \times 26$ the curve is

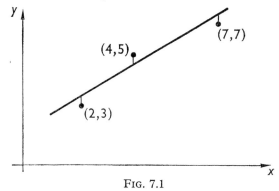

FIG. 7.1

an ellipse; and for different values of d the curves $F(m, c) = d$ are ellipses which differ in size. Standard results of elementary Cartesian geometry show that the directions of their axes are always the same and their centres coincide at the point given by the equations:

$$69m + 13c = 75 \quad \text{and} \quad 13m + 3c = 15. \tag{7.1}$$

These equations have the solution $m = 0.79$ and $c = 1.58$. For large values of d the corresponding ellipses are large, for sufficiently small values of d the ellipses are imaginary, and for a critical value of d the ellipse is a single point—the point just determined. This value of d is the smallest value which $F(m, c)$ can take, and so the required values of m and c have been found. The ellipses are rather like contour lines around a depression.

An alternative method is to differentiate $F(m, c)$ partially with respect to m and c; this gives the same equations (7.1), and the same values of m and c.

The next task is to see how these methods apply in more general cases. The main problem is to see how the numbers in the initial data are modified to give the coefficients in Eqs. (7.1). If the points were co-linear then m and c would satisfy the equations

$$2m + c = 3, \quad 4m + c = 5, \quad 7m + c = 7;$$

but these three equations in two unknowns are inconsistent and no solution can be found. In matrix form we may write in general $\mathbf{Am} = \mathbf{y}$, where in this case

$$\mathbf{A} = \begin{bmatrix} 2 & 1 \\ 4 & 1 \\ 7 & 1 \end{bmatrix}, \quad \mathbf{m} = \begin{bmatrix} m \\ c \end{bmatrix} \text{ and } \mathbf{y} = \begin{bmatrix} 3 \\ 5 \\ 7 \end{bmatrix}.$$

The errors, expressed as the components of a 3-vector, are $\mathbf{Am} - \mathbf{y}$. To minimize the sums of the squares it is necessary to minimize $(\mathbf{Am} - \mathbf{y})'(\mathbf{Am} - \mathbf{y})$.

$$(\mathbf{Am} - \mathbf{y})'(\mathbf{Am} - \mathbf{y}) = (\mathbf{m}'\mathbf{A}' - \mathbf{y}')(\mathbf{Am} - \mathbf{y})$$
$$= \mathbf{m}'\mathbf{A}'\mathbf{Am} - \mathbf{y}'\mathbf{Am} - \mathbf{m}'\mathbf{A}'\mathbf{y} + \mathbf{y}'\mathbf{y}. \quad (7.2)$$

The vector $\mathbf{m}$ therefore has to be chosen to minimize this expression. Now Eqs. (7.1) are in fact the equations

$$\mathbf{A}'\mathbf{Am} = \mathbf{A}'\mathbf{y} \quad (7.3)$$

(verify this), and the solution obtained previously was obtained by solving just these equations. In the general case the same procedure can be followed, because whilst the system $\mathbf{Am} = \mathbf{y}$ has more equations than unknowns, the system (7.3) is square and has just as many equations as there are unknowns. In general such a system of equations has a unique solution, and in order to keep the exposition as simple as possible we will not consider the exceptional cases. It is remarkable that pre-multiplying the 'impossible' system of equations by the matrix $\mathbf{A}'$ converts it to a 'possible' system—and furthermore to a system which is of direct application in the present problem. We will now show that the solution of (7.3) does in fact minimize (7.2).

Note that if (7.3) is satisfied, then, by transposing, $\mathbf{m}'\mathbf{A}'\mathbf{A} = \mathbf{y}'\mathbf{A}$.

Denote the particular value of $\mathbf{m}$ which satisfies Eq. (7.3) by $\mathbf{m}_0$, and the general value of $\mathbf{m}$ by $\mathbf{m}_0 + \boldsymbol{\zeta}$. Substituting this in expression (7.2), after a little reduction the expression becomes

$$\mathbf{y}'\mathbf{y} - \mathbf{m}_0'\mathbf{A}'\mathbf{Am}_0 + \boldsymbol{\zeta}'\mathbf{A}'\mathbf{A}\boldsymbol{\zeta}.$$

Now the first two terms in this expression are constant, and we can see that the third is always positive, unless $\boldsymbol{\zeta} = 0$ when it is zero. It is positive since $\boldsymbol{\zeta}'\mathbf{A}'\mathbf{A}\boldsymbol{\zeta}$ may be written as $\boldsymbol{\eta}'\boldsymbol{\eta}$ where $\boldsymbol{\eta} = \mathbf{A}\boldsymbol{\zeta}$, and $\boldsymbol{\eta}'\boldsymbol{\eta} = \eta_1^2 + \ldots$; that is to say it is the sum of squares and only vanishes if all of its components vanish. But $\boldsymbol{\eta} = 0$ implies that $\boldsymbol{\zeta} = 0$. So to minimize expression (7.2) $\boldsymbol{\zeta}$ must be taken as zero and $\mathbf{m}$ must be taken as $\mathbf{m}_0$, the solution of Eqs. (7.3).

Note: This proof is valid for the cases of practical interest, when $\mathbf{A}$ is of rank 2. If exceptional cases are to be covered also then some details in the above argument require modification.

These matrix considerations have enabled the method to be formulated in general terms, and it may be employed to find the line of best fit through any number of points. The line found by this method is called *the line of regression of y on x* and it provides the best estimate of y for a given value of x. This does not necessarily provide the best estimate of x for given value of y because this is provided by the line of regression of x on y. Those interested may pursue these matters further in more detailed texts.

We will apply the technique to the problem of roasting a turkey. The guide which the manufacturers supply with the gas cooker in my home gives the following instructions:

Weight of bird	Roasting time
5–10 lb	25 min per lb plus 25 min over,
10–14 lb	20 min per lb plus 20 min over,
14–18 lb	16 min per lb plus 16 min over,
18–25 lb	14 min per lb plus 14 min over.

If we draw a graph of cooking time against weight we get four line segments, which have rather surprising discontinuities.

A 10-lb bird requires 275 min on one formula and 220 min on the other. This is quite a difference, and similar discrepancies occur at 14 lb and 18 lb. Is it possible to provide one formula which fits reasonably well over the whole range? This is a good exercise in approximate methods of drawing, and the student can get a good enough fit by drawing in a line by eye. But let us apply the technique described above and try and fit a line to the mid-points of the ranges described by the four rules above. We aim to fit as well as we can the four points (7·5, 212·5), (12, 260), (16, 272) and (21·2, 315). Using these numbers

$$\mathbf{A} = \begin{bmatrix} 7 \cdot 5 & 1 \\ 12 & 1 \\ 16 & 1 \\ 21 \cdot 5 & 1 \end{bmatrix}, \quad \mathbf{y} = \begin{bmatrix} 212 \cdot 5 \\ 260 \\ 272 \\ 315 \end{bmatrix},$$

$$\mathbf{A'Am} = \mathbf{A'y},$$

and so

$$\begin{bmatrix} 918 \cdot 5 & 57 \\ 57 & 4 \end{bmatrix} \begin{bmatrix} m \\ c \end{bmatrix} = \begin{bmatrix} 15838 \\ 1059 \cdot 5 \end{bmatrix}.$$

Solving these equations we get

$$m = 6 \cdot 966, \quad c = 165 \cdot 6.$$

This means that a roasting time of 166 min + 7 min per lb might be reasonable. At the four weights chosen this gives times of 218·5, 250,

278 and 316·5 min, respectively; and these are certainly close enough considering the size of the discontinuities in the manufacturer's formula.

The formula we have derived is clearly most in error for the smallest birds, and there is another possibility which might yield a better result. There are grounds for believing that a formula of the type $y = ax^n$ would be more appropriate. If logarithms are taken

$$\log y = n \log x + \log a,$$

and choosing the best values of n and $\log a$ is a problem of exactly the same type as before. For elementary numerical work it is most common to take logarithms to base 10 for calculations of this kind; but in special circumstances other bases may be more convenient.

It turns out that the curve of best fit is

$$y = 103 \cdot 6 x^{0 \cdot 359},$$

and whilst this is better for small birds than the linear formula (and gives the correct result that a bird of zero weight requires no cooking!), in fact, summing the squares of the errors, it does not fit the four points as well as the straight line does. A very simple and elegant study, using the method of dimensions, by Klamkin [8] shows that in theory the time taken to roast a joint is proportional to the $\frac{2}{3}$ power of its weight. The manufacturers of our gas cooker seem unaware of Klamkin's formula, and they give a rule which is nearer to a $\frac{1}{3}$ power law.

Exercise 7.1

A very good example on logarithmic curve fitting is suggested by Sawyer [10]. The world athletics records for running various distances may be found in a suitable reference book, and if they are plotted on logarithmic graph paper it will be seen that they lie very nearly on a straight line. A good fit can be obtained drawing a line by eye, but of course the exact equation of the line of best fit can be obtained by the methods just described. Sawyer remarks that the time t in seconds is approximately related to the distance s in metres by the formula

$$t^9 = s^{10}/10^{11}.$$

Two things are worth noting when doing this exercise. The points are obviously more reasonably spaced when they are plotted on logarithmic scales than when they are plotted on linear scales. Also, it is noticeable that the linear relation (on logarithmic paper) is astonishingly accurate for distances of 800 metres and above, whereas the points for 100, 200 and 400 metres are somewhat off the line, suggesting in support of common experience that different factors may affect performances in the sprints and the longer distances.

Exercise 7.2

Many examples of curve fitting as it is used in the chemical industry are contained in a monograph by Gregg, Hossell and Richardson[4]. The following formulae concern moving averages, i.e. averages of output or of demand which are taken over fixed periods of years. y_t denotes the value of the variable at year t.

For a 5-year period:

$$5 \text{ (moving average at year } t) = y_{t-2} + y_{t-1} + y_t + y_{t+1} + y_{t+2}.$$
$$10 \text{ (slope at year } t) = -2y_{t-2} - y_{t-1} + y_{t+1} + 2y_{t+2}.$$

For a 7-year period:

$$7 \text{ (moving average at year } t)$$
$$= y_{t-3} + y_{t-2} + y_{t-1} + y_t + y_{t+1} + y_{t+2} + y_{t+3}.$$
$$28 \text{ (slope at year } t) = -3y_{t-3} - 2y_{t-2} - y_{t-1} + y_{t+1} + 2y_{t+2} + 3y_{t+3}.$$

Establish these formulae by finding the line of best fit, and deduce also that for a 5-year period:

$$10 \text{ (slope at year } t) = 10 \text{ (slope at year } t-1) + 3y_{t+2} + 2y_{t-3} -$$
$$- 5 \text{ (moving average at year } t),$$

and for a 7-year period:

$$28 \text{ (slope at year } t) = 28 \text{ (slope at year } t-1) + 4y_{t+3} + 3y_{t-4} -$$
$$- 7 \text{ (moving average at year } t).$$

7.2 Multiple regression analysis

Similar methods may be employed to obtain linear equations of best fit when one variable (the dependent variable) is known to be a function of a number of others (the independent variables). There is a certain discrepancy between statistical terminology and terminology in other parts of mathematics at this point, and as we are more concerned with general mathematical ideas, and less with the details of statistical practice, we are using *dependent* and *independent* in the way that is common practice in elementary courses. We might wish to think of the heights of male children as being, to some extent, dependent on the two variables, height of father and height of mother, and wish to determine m_1 and m_2 so that an equation

$$y = m_1 x_1 + m_2 x_2, \qquad (7.4)$$

where y = height of male child, x_1 = height of father and x_2 = height of mother, fits collected data as closely as possible. x_1 and x_2 are independent variables in the usual sense of school geometry and calculus, but in a statistical sense they may be interconnected (correlated, interdependent), and it runs counter to statistical practice to call variables 'independent' when they are in fact 'interdependent'. This example has been chosen because we can see a reason why the heights of mothers and fathers are not, statistically speaking, independent. There is some tendency for tall men and women to marry one another and for short men and women to marry one another, and this comes out as a statistical correlation between the variables x_1 and x_2 if a number of measurements are taken. It is common statistical practice to work about mean values, so that there is no constant term in Eq. (7.4). In the problem of choosing the best constants, m_1 and m_2, to fit the data it is beside the point whether the variables x_1 and x_2 are correlated or not. We assume that we have a number of measurements of the three variables in question, that we have sets of values $(x_1^{(1)}, x_2^{(1)}, y^{(1)})$, $(x_1^{(2)}, x_2^{(2)}, y^{(2)})$, etc., the superscript referring to the individual on which the three measurements are made. We will assume that there are n individuals. Proceeding as before, this time we wish to choose m_1 and m_2 to minimize

$$S = \sum_{r=1}^{n} \left(y^{(r)} - m_1 x_1^{(r)} - m_2 x_2^{(r)} \right)^2.$$

Differentiating, or alternatively employing the earlier geometrical argument,

$$\frac{\partial S}{\partial m_1} = -2 \sum_{r=1}^{n} x_1^{(r)} \left(y^{(r)} - m_1 x_1^{(r)} - m_2 x_2^{(r)} \right)$$

and

$$\frac{\partial S}{\partial m_2} = -2 \sum_{r=1}^{n} x_2^{(r)} \left(y^{(r)} - m_1 x_1^{(r)} - m_2 x_2^{(r)} \right).$$

Equating these expressions to zero in order to get the critical values of m_1 and m_2, we get

$$\sum_{r=1}^{n} x_1^{(r)} \left(y^{(r)} - m_1 x_1^{(r)} - m_2 x_2^{(r)} \right) = 0$$

and

$$\sum_{r=1}^{n} x_2^{(r)} \left(y^{(r)} - m_1 x_1^{(r)} - m_2 x_2^{(r)} \right) = 0.$$

Denoting the sets of measurements of x_1, x_2 and y respectively by vectors $\mathbf{x}_1$, $\mathbf{x}_2$ and $\mathbf{y}$ these equations can be written in matrix form,

$$\mathbf{x}'_1 (\mathbf{y} - m_1 \mathbf{x}_1 - m_2 \mathbf{x}_2) = 0,$$
$$\mathbf{x}'_2 (\mathbf{y} - m_1 \mathbf{x}_1 - m_2 \mathbf{x}_2) = 0.$$

Matrix notation permits even further economies, because using block matrices these equations become

$$\begin{bmatrix} \mathbf{x}_1' \\ \cdots \\ \mathbf{x}_2' \end{bmatrix} \mathbf{y} = \begin{bmatrix} \mathbf{x}_1' \\ \cdots \\ \mathbf{x}_2' \end{bmatrix} [\mathbf{x}_1 \mid \mathbf{x}_2] \begin{bmatrix} m_1 \\ m_2 \end{bmatrix}.$$

$$2 \times n \qquad n \times 1 \quad 2 \times n \qquad n \times 2 \quad 2 \times 1$$

As a check the sizes of the matrices are printed underneath. Matrix notation is achieving a very substantial economy at this point, and the student is advised to write the equations out in full until the significance of the contracted notation is appreciated.

If we denote the matrix of the assembled measurements of the x_1 and x_2 values by

$$\mathbf{X} = [\mathbf{x}_1 \mid \mathbf{x}_2],$$

and denote the column vector $(m_1, m_2)'$ by $\mathbf{m}$, then the equations determining m_1 and m_2 become

$$\mathbf{X}'\mathbf{y} = \mathbf{X}'\mathbf{X}\mathbf{m}.$$

Hence
$$\mathbf{m} = (\mathbf{X}'\mathbf{X})^{-1}\mathbf{X}'\mathbf{y}. \tag{7.5}$$

This formula, which is essentially equation (7.3) of the previous section, gives the *regression coefficients* m_1 and m_2 explicitly in terms of the experimental measurements of x_1, x_2 and y. It is well worth the labour of getting it in this form, because now it applies to cases where y is a function of any number of variables $x_1, x_2, \ldots$. This illustrates one of the great advantages of matrix notation; work in two variables that is properly formulated can often be extended to any number of variables just as easily. In passing we may refer back to Exercise 5.1, and note that the way in which the matrix $\mathbf{X}$ is involved in Eq. (7.5) reminds us of projection operators.

For examples of this kind of statistical technique in an educational context we may refer to the Plowden Report [15], entitled *Children and their Primary Schools*. The second volume of this report contains extensive statistical appendices and is a mine of information both on educational matters and also on current statistical practice. It shows among other things the very large numbers of variables which may be considered when modern computers are employed.

In such enquiries a regression analysis is carried out by fitting data to a sequence of equations

$$y = m_1 x_1,$$
$$y = m_1' x_1 + m_2' x_2,$$
$$y = m_1'' x_1 + m_2'' x_2 + m_3'' x_3, \text{ etc.,}$$

adding one more variable each time. The order in which extra variables are taken into account is to some extent a matter of choice, some arbitrary criterion being selected, but in any case the expertise of the statistician is heavily involved. In practice in an enquiry of this nature the process tends to 'settle down', and it becomes apparent that the variations of the variable y are adequately explained in terms of a certain number of variables x, and nothing is gained by including more.

In Appendix 4 of the report, 'The regression analyses of the National Survey', by G. F. Peaker, the techniques of regression analysis are described and also its application to a national survey the object of which was to test conjectures about the reasons why some children make more progress in school than others. Two kinds of analysis were attempted, one comparing different schools and the other comparing different pupils within the same school. The initial lists of variables, the many things which might be expected to influence a child's educational attainment, contained 104 variables for the analyses between schools and 73 for the analyses between pupils within schools. Initial regression analyses showed in each case that about a dozen variables made significant contributions to the criterion variation.

After further analysis seven of these variables were finally taken as significant, but of these seven one could be seen to be very much the most important. This especially important factor was the factor of parental attitudes, the concern of the parents for the progress of their child. It may be said that this is only what was to be expected, but the statistical analysis brought out very clearly the extent to which this factor predominates. The analysis showed that the variation in parental encouragement and support has much greater effect than either the variation in home circumstances or the variation in schools.

The analysis also pointed to a further conclusion with considerable social implications. 'Although the variation in parental material circumstances and parental education can account for some of the variation in parental attitudes, it cannot account for very much, and leaves open the possibility that attitudes may be changed by persuasion.' Since this was written attempts have been made to pursue its implications.

An interesting side issue in this enquiry was the effect of class size. It is something of an embarrassment to educational statisticians that their attempts to demonstrate that large classes are detrimental to children's progress meet with failure. This, it seems, is one thing which statisticians cannot get figures to prove! The Plowden Report says about this, 'It is very hard to believe that, if other things are equal, merely adding several more children to the class will improve the average achievement. Yet most surveys, including our own, show mild positive simple correlations between the average size of class in a school and the average test score for that school.' They hasten to point out that this is

7.3 Buns, cakes and principal components

not the whole story and to suggest other factors which account for this, at first sight, puzzling phenomenon. One such factor is that remedial groups in schools are usually made small deliberately.

7.3 Buns, cakes and principal components

We may start with some problems of a type which are now standard in many of the new mathematics programmes in schools. To make buns and cakes we require the following ingredients (fill in suitable quantities for yourself!).

		Bun	Cake
A	Flour	—	—
	Sugar	—	—
	Butter	—	—
	Currants	—	—
	Milk	—	—

Alice, Barbara and Doreen are going to make the following numbers of buns and cakes (insert your own numbers),

		Alice	Barbara	Doreen
B	Buns	—	—	—
	Cakes	—	—	—

It is easy to see that the amounts of ingredients which each girl will need are given by a matrix multiplication.

		Alice	Barbara	Doreen
M = AB	Flour	—	—	—
	Sugar	—	—	—
	Butter	—	—	—
	Currants	—	—	—
	Milk	—	—	—

Now we might feel that the numbers in the matrix **M** should reveal in some way that it is just two recipes which they are making up rather than one or three or any other number. Putting it another way, given only **M** can we make any deductions about what is happening?

There is no unique factorization of matrices, so it is hopeless to imagine that we can get **A** and **B** separately merely from a knowledge of **M**.

We may think of things geometrically by adopting a co-ordinate system with axes for flour, sugar, ... , milk and plotting points whose co-ordinates are equal to the amounts of those materials used, one for each girl. This requires a five-dimensional space, but a sketch as shown in Fig. 7.2 will do.

Instead of thinking of a girl as being represented by a point we may think instead of the vector from the origin to the point. What can we say about the girl vectors? Instead of being spread out in all directions they lie in a plane, because the materials which each girl uses are due simply to so many buns and so many cakes. Each girl vector is in the

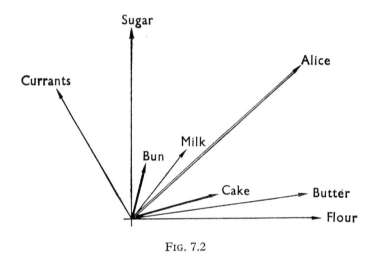

Fig. 7.2

plane determined by the bun and cake vectors. The same thing would apply if a whole class of girls made up just the two recipes.

Given the matrix **M** how can we decide that its columns lie in a two-dimensional subspace of the ingredient space? This is a problem about the rank of a matrix, and any theorems on rank which the reader may know apply, but we will not assume any of these here. One answer can be got as follows.

It is useful to extend the geometrical picture. We may envisage a large number of girls engaged in cookery, and we may assume that there are so many 'like Alice', so many 'like Barbara' and so on. 'Like' means that each makes just as many buns as Alice makes and just as many cakes. There is now a set number of types of girl, girls like Alice, Barbara, ... , Doreen.

A class of girls is then described by a vector,

$$\mathbf{x} = \begin{bmatrix} \text{Number of girls like Alice} \\ \text{Number of girls like Barbara} \\ \cdot \quad \cdot \quad \cdot \quad \cdot \quad \cdot \\ \text{Number of girls like Doreen} \end{bmatrix}.$$

All vectors like this form the x-space. Then, increasing the number of columns in the matrix $\mathbf{B}$ to include all the various types of girl,

$$\mathbf{Bx} = \begin{bmatrix} \text{Number of buns made} \\ \text{Number of cakes made} \end{bmatrix}.$$

Then we may put

$$\mathbf{y} = \mathbf{ABx} = \mathbf{Mx} = \begin{bmatrix} \text{Total flour used} \\ \cdot \quad \cdot \quad \cdot \\ \text{Total milk used} \end{bmatrix},$$

and we see that we are mapping the x-space into the y-space, and all the previous geometrical ideas of mappings apply. In particular the number of recipes being made up corresponds to the number of dimensions of the image of the x-space in the y-space. This is two in our particular example.

Now the dimension of the image can be found by considering the image of the unit sphere in the x-space, as in Section 2.9. The geometrical considerations apply; the unit sphere in the x-space is transformed to an ellipsoid and the number of non-zero semi-axes is equal to the number of non-zero eigenvalues of $\mathbf{M'M}$. In this particular case the ellipsoid is a two-dimensional ellipse, the curve together with its interior. Note that the type of analysis now being described can decide that the girls are making up two recipes, but it cannot decide what the recipes are individually, it can only identify the subspace which the recipes span.

In the practical applications of these techniques further considerations are very important. Imagine that the girls are doing an examination and as part of the test they have to remember their quantities. Now their vectors will vary, and they will depart a little from the plane in the ingredient space where they were all situated before. The elliptic disc which contained the girl-vectors will not be flat, but will be spread a little.

If we inspected the matrix $\mathbf{M}$ now we could not conclude that the girls were using the ingredients they were because they were making up two recipes, we would want techniques which enabled us to conclude that there were probably two recipes, although there was some error. The use of the eigenvalues of $\mathbf{M'M}$ would enable us to do this, as we would have two eigenvalues which were clearly non-zero and some more which were 'small'. We would have to use our judgement, and decide to accept the two larger eigenvalues as 'significant' and to attribute the 'small' ones to random fluctuations.

Exercise 7.3

Consider what happens if buns and cakes use ingredients in exactly the same proportions, the differences being in the cooking. The image of the x-space in the y-space is then of dimension 1, and it is as if they were only making up one recipe.

The above example is very unsophisticated from the point of view of statistical refinements, but it serves to introduce a number of the basic ideas. Before going on perhaps one further feature might be noted. If two girls are making buns and cakes in very much the same proportion then their vectors in the ingredient space are comparatively near together, as far as their directions are concerned. If they are making buns and cakes in very different proportions then their vectors are inclined at a larger angle to one another. Later developments of the technique refine this idea and ways are devised of making this angle a precise measure of a statistically significant parameter.

In the above discussion it was purely a convention to set up an ingredient space and to plot the girl-vectors in it. We started with a matrix and constructed a space in which the number of dimensions (co-ordinates) was equal to the number of rows of the matrix, and each 'point' plotted, that is each girl-vector, corresponded to a column of the matrix.

From an algebraic point of view a matrix is a set of numbers and it may be analysed by rows or by columns. It is just as reasonable, mathematically speaking, to construct a space in which the co-ordinate axes correspond to the girls and the points correspond to the rows of the matrix, that is to the ingredients. In this particular problem this formulation does not seem so instructive, but in other problems it can be the preferable point of view. If we work this way we are led eventually to consider the eigenvalues of **MM'**, instead of **M'M**.

It is a piece of pure mathematics to show that **MM'** and **M'M** always have the same non-zero eigenvalues and so we are led to precisely the same conclusions about dimension as before—in terms of our problem the conclusion that *just two* items were being cooked. **MM'** and **M'M** are both square, but in general they differ in size, and in general they will have a different number of eigenvalues; this is accounted for by the fact that they may have zero eigenvalues of different multiplicities.

The techniques of factor analysis were developed in the first place around problems of educational testing, especially in connexion with intelligence testing and educational selection. This being so it will be convenient to formulate the next problem for discussion in these terms.

BUNS, CAKES AND PRINCIPAL COMPONENTS

As a result of subjecting a class of pupils to a battery of tests, or examinations, we end up with a mark sheet drawn up in the familiar way.

	Test 1	Test 2	.	.
Pupil 1	.	.	.	.
Pupil 2	.	.	.	.
Pupil 3	.	.	.	.
.	.	.	.	.
.	.	.	.	.
.	.	.	.	.

We will call this mark matrix **M**. (Some authors work with the transpose of this matrix, but that is merely a superficial difference, it does not affect the ideas involved.)

For statistical purposes it is usual to normalize the columns of this matrix; this means simply that, as is frequently the case in school examinations, the marks are scaled in a particular way. For statistical purposes the marks of each test are adjusted so that their mean is zero and the sum of their squares is one. This involves first subtracting the mean mark from every mark, and then dividing all through by the square root of the sum of the squares. This is standard practice in statistics and we will assume that the columns of the **M** we are working with have been so normalized, although most of the subsequent analysis can be carried out in the same way, whether the initial data has been normalized or not. This process of normalization is exactly the same as the conversion of direction ratios to direction cosines, described at the end of Section 2.6.

The assumption is now made that certain basic abilities, or *factors*, are being tested, that the particular tests involve these factors to varying extents, and that the pupils perform as they do because they possess the requisite abilities to varying degrees. As we are working now

- the pupils correspond to ingredients in the previous example;
- the tests correspond to pupils in the previous example;
- the factors, which are unknown and have to be found, correspond to the buns and the cakes in the previous example.

(That pupils do not correspond to pupils in the two examples should not be an obstacle. We have said that the first example can be analysed either way round, also that practising statisticians differ in whether their standard matrix is **M** or **M'**.)

We will assume that there are n pupils and t tests, and that there are f factors. It is essential to realize that we are working on the assumption that there are these things called 'factors' to be found. This hypothesis is fundamental to the method. Human abilities are less susceptible to precise numerical description than are buns and cakes. It is quite possible to question the whole basis on which the techniques of factor analysis rest, but this is no place for such a discussion; we are merely trying to show how the mathematics works if the experimenter decides to employ the method.

The number of factors at work is equal to the rank of the matrix **M**, but we will aim to make the discussion independent of the notion of rank because some readers may not work easily with it. It is not difficult to prove that **M**, **M'M** and **MM'** all necessarily have the same rank; but once again this is a result to which we will not need to appeal. Corresponding to each factor there is a non-zero eigenvalue of **M'M**. In practice, because of the statistical fluctuations in the marks, **M'M** has a range of eigenvalues and the experimenter has to decide that some of these are so small as to be negligible. Which eigenvalues are to be treated as zero and which are to be retained is a matter for judgement.

As well as deciding on the number of factors which are operating the experimenter also wishes to decide to what extent the different tests involve each factor, and to what extent the pupils have the corresponding abilities. In the introductory example the corresponding things were the extent to which the different ingredients were involved when buns and cakes were made.

The basic problem is therefore to express the matrix **M** in the form

$$\mathbf{M} = \mathbf{ST},$$

where **M** is an $n \times t$ matrix, giving the marks obtained by each of the n pupils in the t tests;

S is an $n \times f$ matrix, giving the weights of the pupils in each factor, i.e. it indicates the extent to which the pupils possess the various abilities;

T is an $f \times t$ matrix, giving the loadings of the different tests on each factor, i.e. it shows the extent to which each test tests the various factors.

This corresponds to the form **M** = **AB** in the introductory example.

There is no sense in which there is unique factorization for matrices, and so before precise answers can be given more conditions have to be imposed. The psychologist usually wants the columns of **S** to be normalized and mutually orthogonal. This means that the factors can be used as an orthogonal co-ordinate system in terms of which pupils and tests can both be expressed.

Exercise 7.4
Show that this requirement amounts to demanding that

$$S'S = I_f,$$

where I_f is the unit $f \times f$ matrix.

Show that this is satisfied if and only if S consists of the first f columns of an orthogonal matrix.

Show also that this does not, in general, imply that SS' is equal to the $n \times n$ unit matrix.

If $M = ST$, then $M'M = T'S'ST$. Now $M'M$ is always symmetric (by Exercise 2.34), and the standard spectral resolution of a symmetric matrix may play its part. That is to say we may utilize the geometrical ideas either of Theorem 2.4 or of Section 5.1. Before doing this let us consider the entries which appear in $M'M$ and their experimental significance. The columns of M are now normalized; that is, they are unit vectors. Position (p, q) of $M'M$ contains the inner product of column p of M with column q. But the columns being normalized this inner product is simply the cosine of the angle between the column vectors. In statistical studies of this kind this cosine is called the *correlation coefficient* between the vectors. In this case the correlation is between the *tests*; the matrix MM' gives the correlations between the pupils. Both of these are important according to the questions asked.

We may ask 'Were the examination results in geography rather like those in history or not?' 'Do the science marks correlate more closely with the marks in English or in mathematics?' The elements of the matrix $M'M$ give statistical answers to those questions. The closer the entry is to unity the closer the correlation between the subjects. Of course, every diagonal element is unity.

We may also ask 'Were Alice's examination results more like Barbara's or Doreen's?' Statistical answers to questions like this appear in the matrix MM'.

The spectral decomposition of $M'M$ can now be used to determine S and T. By Theorem 2.4

$$M'M = H'DH,$$

where D is a diagonal matrix having as diagonal elements the eigenvalues $\lambda_1, \lambda_2, \ldots,$ of $M'M$. H is an orthogonal matrix which has the normalized eigenvectors of $M'M$ as *rows*.

The novelty here is that many of the λ_r are zero or effectively zero. Because of the statistical variations in the marks it is to be anticipated that a number of the λ_r will be small, without actually vanishing, and a decision has to be taken as to how many λ_r should be retained and how many should be ignored. In practice the λ_r have to be computed successively, starting with the largest, but in order to see the overall picture we will assume here that they are all known at this stage. We assume that $\lambda_1 > \lambda_2 > \ldots > \lambda_f > 0$; and after this point the eigenvalues are all effectively zero. (It may be proved that none of the eigenvalues of $\mathbf{M'M}$ are negative when, as here, all the elements of $\mathbf{M}$ are real.)

Then
$$\mathbf{M'M} = \mathbf{H'} \begin{bmatrix} \lambda_1 & & & 0 \\ & \ddots & & \\ & & \lambda_f & \\ 0 & & & 0 \end{bmatrix} \mathbf{H},$$

$$= \mathbf{H'} \begin{bmatrix} \sqrt{(\lambda_1)} & & \\ & \ddots & \\ & & \sqrt{(\lambda_f)} \\ & 0 & \end{bmatrix} \begin{bmatrix} \sqrt{(\lambda_1)} & & & \vdots \\ & \ddots & & \vdots 0 \\ & & \sqrt{(\lambda_f)} & \vdots \end{bmatrix} \mathbf{H},$$

where the orders of the matrices on the right-hand side are $t \times t$, $t \times f$, $f \times t$ and $t \times t$, respectively.

Now define $\mathbf{T}$ by
$$\mathbf{T} = \begin{bmatrix} \sqrt{\lambda_1} & & & \vdots \\ & \ddots & & \vdots 0 \\ & & \sqrt{\lambda_f} & \vdots \end{bmatrix} \mathbf{H};$$

then, as will be shown, this matrix is satisfactory as $\mathbf{T}$, and

$$\mathbf{M'M} = \mathbf{T'T}.$$

$\mathbf{T}$ can be expressed a little more simply as
$$\mathbf{T} = \mathbf{\Lambda}^{\frac{1}{2}} \mathbf{H}_f,$$

where $\mathbf{\Lambda}^{\frac{1}{2}} = \text{diag}\{\sqrt{\lambda_1}, \sqrt{\lambda_2}, \ldots, \sqrt{\lambda_f}\}$
and H_f is the matrix consisting of the first f rows of $\mathbf{H}$ only.

We still wish to find a matrix $\mathbf{S}$, such that
$$\mathbf{M} = \mathbf{ST} = \mathbf{S}\mathbf{\Lambda}^{\frac{1}{2}} \mathbf{H}_f.$$

Post-multiplying this equation by $\mathbf{H}'_f \mathbf{H}_f$ we have
$$\mathbf{MH}'_f \mathbf{H}_f = \mathbf{S}\mathbf{\Lambda}^{\frac{1}{2}} \mathbf{H}_f \mathbf{H}'_f \mathbf{H}_f.$$

It is easy to show that $\mathbf{H}_f \mathbf{H}'_f = \mathbf{I}_f$ (prove this), and so
$$\mathbf{MH}'_f \mathbf{H}_f = \mathbf{S}\mathbf{\Lambda}^{\frac{1}{2}} \mathbf{H}_f = \mathbf{M}.$$

Observe this result carefully; and note that although it is *not* true that $\mathbf{H}'_f \mathbf{H}_f = \mathbf{I}_t$, in spite of this it is nevertheless true that $\mathbf{MH}'_f \mathbf{H}_f = \mathbf{M}$.

Interchanging the right- and left-hand sides we may proceed,

$$\mathbf{M} = \mathbf{MH}'_f\mathbf{H}_f = \mathbf{MH}'_f\mathbf{I}_f\mathbf{H}_f,$$
$$= \mathbf{MH}'_f\Lambda^{-\frac{1}{2}}\, \Lambda^{\frac{1}{2}}\mathbf{H}_f$$

since the inserted diagonal matrices have $\mathbf{I}_f$ as their product.

The expression on the right-hand side may now be split as required. The last two factors have $\mathbf{T}$ as their product, so we may define $\mathbf{S}$ by

$$\mathbf{S} = \mathbf{MH}'_f\Lambda^{-\frac{1}{2}},$$

and we have the required relation

$$\mathbf{M} = \mathbf{ST}.$$

Exercise 7.5

Confirm that $\mathbf{S}$ has the required property, $\mathbf{S}'\mathbf{S} = \mathbf{I}_f$.

The process above, which was devised by Hotelling [6] in 1933, is standard practice in factor analysis; but because of the size of practical problems the spectral decomposition has to be done in practice by proceeding a stage at a time, much as with the process for locating eigenvectors described in Section 2.9, or with the methods at the end of Section 6.6. Details of practical methods may be found in suitably specialized texts.

This process, using the standard spectral decomposition of $\mathbf{M}'\mathbf{M}$, separates out the parts played by the various factors involved. We have

$$\mathbf{M}'\mathbf{M} = \mathbf{T}'\mathbf{T} = \mathbf{H}'_f\Lambda\mathbf{H}_f = \Sigma\, \lambda_r\mathbf{h}'_r\mathbf{h}_r,$$

where $\mathbf{h}_r$ is the normalized *row* vector corresponding to the eigenvalue λ_r. The matrix of weights $\mathbf{S}$ and the matrix of loadings $\mathbf{T}$ may be decomposed in a corresponding manner. Writing them out more fully,

$$\mathbf{S} = \mathbf{M}\left[\frac{1}{\sqrt{\lambda_1}}\mathbf{h}'_1, \ldots, \frac{1}{\sqrt{\lambda_f}}\mathbf{h}'_f\right]$$

and

$$\mathbf{T} = \begin{bmatrix} \sqrt{\lambda_1}\mathbf{h}_1 \\ \cdot \\ \cdot \\ \cdot \\ \sqrt{\lambda_f}\mathbf{h}_f \end{bmatrix}.$$

This shows how each factor is associated with a column of $\mathbf{S}$ and with a row of $\mathbf{T}$.

This method of analysis is called *principal component analysis*. It may be as well to emphasize that this method is used to identify the factors which are acting, and as we have seen they are identified by algebraic calculations which are closely related to the calculations which are used to find the principal axes of quadric surfaces. This method is not appropriate to such statistical tasks as measuring the influence of something which the experimenter has previously decided is a factor on other variables in the enquiry. For this type of job a regression analysis is used.

7.4 Some geographical problems

The application of these techniques to some geographical problems will now be described. Principal component analysis has been widely used in geography, and the examples quoted below are by no means isolated cases, nor by any means the earliest.

The first example summarizes a study of the geographical regions of the United States of America by Cole and King [3]. Data on 25 geographical variables for the 50 states of the USA was obtained from the *Statistical Abstract of the United States*, 1965. The 50 states are, as it were, 50 'pupils' being subjected to 25 'tests'. Many geographical variables could have been chosen, but those selected were:

 (i) three on physical features; altitude, temperature, precipitation;
 (ii) eight related to population; percentage of urban population, birth rate, etc.;
 (iii) eleven economic–social variables; income, number of automobiles per thousand inhabitants, etc.;
 (iv) three electoral variables; percentage of Republican votes at three different elections.

These variables were tabulated as a matrix $\mathbf{M}$, the columns being normalized as described previously, and the correlation matrix $\mathbf{M'M}$ was then calculated. The largest eigenvalue of the 25×25 matrix was 8·74. What was the corresponding factor? The eigenvector describing this factor turned out to be related most closely to the quality of housing (as measured by the percentage of houses with all plumbing facilities sound) where there was a loading (direction cosine) of 0·94, it was closely related also to the proportion of people having telephones, to income and to the proportion of the population which was urban. It pointed therefore to the contrast between the wealthier, more heavily urbanized areas in the north-east and on the west coast on the one hand and the area of the deep south on the other.

The second factor, with eigenvalue 4·58, correlated positively with altitude and negatively with manufacturing and precipitation. A map showing the areas high and low in the factor shows two very clear regions, the desert and mountain regions of the west contrasted with the east coast.

The third factor, with an eigenvalue of 3·14, correlated more closely with population change, migration and birth rate than with anything else, and this contrasted the extreme south-east and south-west on the one hand with the central states on the other.

When these three classifications are mapped they can be seen to correspond to sensible comparisons which a geographer would accept, and to be significantly different methods of comparison.

The next two eigenvalues were 1·78 and 1·00.

It is useful to be able to estimate how much of the total variance the factors associated with the different eigenvectors actually explain. Since the correlation matrix $M'M$ has ones in every position on the principal diagonal the total of the entries on the diagonal is always equal to the order of the matrix—in this case 25. In any matrix the total of the terms on the principal diagonal is equal to the sum of the eigenvalues. Therefore in this example the eigenvalues total 25.

It is plausible that the amount of variation which the factor corresponding to an eigenvalue explains is measured by the relative importance of the eigenvalue, that is by its ratio to the total of the eigenvalues. In this case the first eigenvalue, 8·74, is 35% of the total, the second, 4·58, 18% of the total, and the third, 3·14, 13% of the total. The three factors might therefore be expected to account for something like two-thirds of the variations being studied. This notion can be made precise by defining a property called *variance*, and it can be shown that variance is measured in this way by the eigenvalues.

This, incidentally, is one of the reasons why the columns of M are normalized before the calculation is undertaken. Scaling the entries in M alters the eigenvalues of $M'M$, and no significance can be attached to the actual size of the eigenvalues unless the entries in M are standardized in some way. In the very first example about buns and cakes this was not done, and therefore there would be no significance in the actual size of the eigenvalues. Note also that in that example the eigenvalues would be altered by altering the units in which the quantities were measured. Thus, the milk might be measured by weight in the same unit as the other ingredients, but again it might be measured by volume in some arbitrarily different units. Questions of this kind have to be resolved if principal component analysis is to be a reliable technique.

This may be a suitable place at which to stress also that this kind of statistical analysis demonstrates *correlations*. The existence of a correlation does not demonstrate cause and effect. Statistical techniques always call for expert interpretation, and the input data must be carefully selected and prepared if it is to give trustworthy results.

Another investigation by King, concerning areas in the north of England, is of interest. Six areas were chosen—the Askrigg Block, the

Alston Block, the Cheviots, the Lake District, the Solway Lowlands and the Northumberland Coast—and twelve geographical variables—such as absolute relief, relative relief, the number of streams, waterfalls and lakes, etc. All the information was read from standard one-inch Ordnance Survey maps, indicating how feasible it might be even for a school to undertake a similar enquiry if they could draw on adequate computing facilities. The correlation coefficients and the factors which emerged from the calculations related well to the non-quantitative judgements which a geographer might make about these regions. Details of this investigation may also be found in Reference [3].

7.5 Miscellaneous examples

Here are some more examples of this type of statistical analysis, which have been chosen for their general interest. The work of the writers quoted does not always employ exactly the techniques which have just been described, but they are the product of similar ways of thinking.

The regression analysis in the Plowden Report has been described earlier: the second volume of this same report [15] also contains, in Appendix 9, an account of a survey of ten-year-old children in the Manchester area in which extensive use was made of principal component analysis. The report gives a great deal of mathematical detail. The enquiry involved 87 variables, but it was found that nearly all the variation was concentrated in the first six factors. (In this enquiry after the factors had been found they were subjected to an additional statistical refinement known as 'rotation'. As the name indicates this notion can be explained in geometrical terms, but we cannot go into the statistical aspects of rotation here.) In fact it was possible to go further than this and say that in nearly all of the separate parts of the Manchester survey, when a principal component analysis was undertaken the first factor alone greatly outweighed all the other factors put together. On a purely commonsense appraisal one might have expected one factor to predominate, but commonsense alone could not anticipate the extent to which the main factor turned out to be dominant.

As a result of the survey the authors drew two conclusions of special importance. 'Environmental forces bear most heavily on the brightest of our children; and the factors in the home are overwhelmingly more powerful than those of the neighbourhood and the school—and of these, factors of parental attitude to education, to the school and to books are of far greater significance than social class and occupational level.' The link with the national survey regression analysis is clear.

A system for classifying human physiques was described in 1940 by Sheldon and his associates [11]. They showed that when a number of body measurements were made on a variety of men the correlation between them was such that the men could be classified into three physical types, *endomorphs, mesomorphs* and *ectomorphs*. This classification is by shape and it ignores overall differences in size. People have the three factors of endomorphy, mesomorphy and ectomorphy in varying proportions. They can be used rather as areal or trilinear co-ordinates are used in plane geometry so that different types of

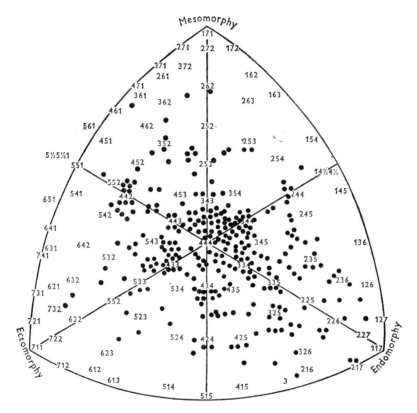

FIG. 7.3 Somatotype distribution of 283 Oxford University students, 1948–50. *Source*: J. M. Tanner, *The Physique of the Olympic Athlete*, Allen & Unwin (1964).

physique can be charted on a two-dimensional, roughly triangular diagram. At the corners of the corner are found the extreme types for each of the three factors.

The extreme endomorph is well rounded, with much fat in the upper arm and thigh, but slender wrists and ankles and rather weak arms and legs. The extreme mesomorph is the classical Hercules; predominantly bone and muscle. The extreme ectomorph is thin, narrow and spindly.

Tanner [13] has made a fascinating study of Olympic athletes, using this classification. This book assembles a great wealth of statistical information about Olympic athletes and shows where the athletes excelling at the different Olympic events cluster on these triangular classification charts.

The four diagrams from this book reproduced here as Figs. 7.3–6 show very vividly the different distributions of physique which occur

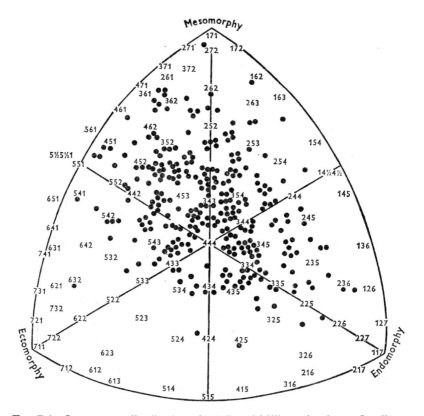

FIG. 7.4 Somatotype distribution of 287 Royal Military Academy, Sandhurst, cadets, 1952. *Source*: J. M. Tanner, *The Physique of the Olympic Athlete*, Allen & Unwin (1964).

in populations of Oxford University students, students at the Royal Military Academy, Sandhurst, students of physical education at Loughborough College of Education, and Olympic track and field athletes.

The methods of principal component analysis have been applied in a wide variety of disciplines—chemistry [5], education [2, 14], biology, geography, medicine and the social sciences. They may be applied to the design of ready-made clothes or to the analysis of stone age axes [1]. The whole problem of classification, as in taxonomy, can be approached by these and by related techniques. An interesting popular account of the use of computers to draw up the family trees of fossil horses and a purely imaginary set of animals has been written by Sokal [12]. Listeners to Alan Lomax's radio programmes of folk songs may be interested to learn that a principal component analysis has been carried out on a

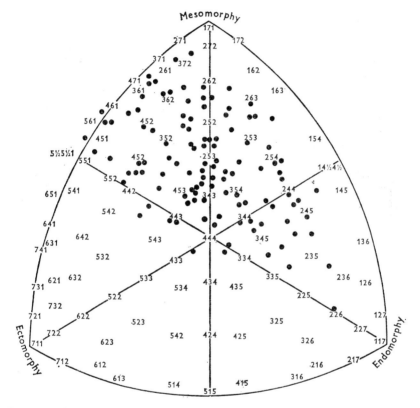

FIG. 7.5 Somatotype distribution of 114 Loughborough College of Education students. *Source*: J. M. Tanner, *The Physique of the Olympic Athlete*, Allen & Unwin (1964).

59-row matrix of folk song statistics and ten factors extracted, of which two accounted for nearly half the total variance [9].

The widespread applicability of the various types of factor analysis is all the more surprising as it proceeds in defiance of what was for many years an established precept in scientific investigation—that it is a desirable strategy to design experiments so that one variable may be changed at a time.

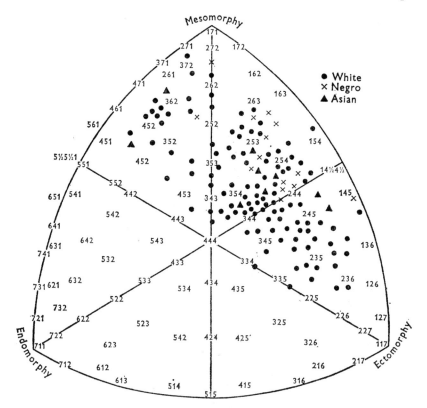

FIG. 7.6 Somatotype distribution of 137 Olympic track and field athletes. *Source*: J. M. Tanner, *The Physique of the Olympic Athlete*, Allen & Unwin (1964).

7.6 An alternative approach to principal components

In this section we give an alternative, perhaps more common, approach to principal components and the covariance matrix. As an introduction we will first consider a numerical example concerning a line of best fit, where the fit is obtained by a new method which should be compared and contrasted very carefully with the method of Section 7.1. This example involves a space of only two dimensions, but afterwards the general problem is considered in a space of t dimensions. Throughout the discussion we draw very heavily on the geometrical ideas of Sections 2.8 and 2.9.

The following method of determining best fit might be appropriate in cases where one variable is expected to be proportional to another. This might occur, for example, if we are concerned with the load on a spring and the extension of the spring. It might also occur in some far less deterministic situations, as when, for example, we consider measurements of the heights and weights of a number of men, and seek a linear relation between departures of height and weight from their respective means.

Let us suppose that we are seeking to establish a relation between variables x_1 and x_2 of the form $x_2 = kx_1$. To start with it is helpful to put this in a form in which the two variables occur symmetrically, and a form which it is easier to generalize to more variables subsequently. Hence we will seek to establish a relation of the form

$$\frac{x_1}{l_1} = \frac{x_2}{l_2}. \qquad (7.6)$$

l_1 and l_2 are the direction cosines of the line, as in Section 2.6.

Assume now that we have three pairs of readings of x_1 and x_2. In practice we would want more, but we will reduce the arithmetic in the example. Let the pairs of readings be (2, 3), (3, 4·8) and (3·8, 6). The problem is to choose l_1 and l_2 in Eq. (7.6) in such a way as to fit the data best. The criterion which we adopt is indicated in Fig. 7.7. The three

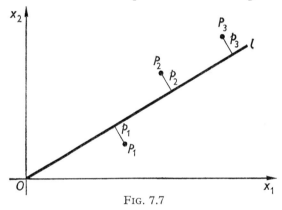

FIG. 7.7

points P_1, P_2, P_3 representing the pairs of measurements are located at perpendicular distances p_1, p_2, p_3 from the desired line, l. Let us agree that the line fits best when $p_1^2 + p_2^2 + p_3^2$ is as small as possible. Compare this very carefully with the situation in Fig. 7.1.

The projection of OP_1 on the line l is, by Eq. (2.24), $2l_1 + 3l_2$. Hence

$$p_1^2 = OP_1^2 - (2l_1 + 3l_2)^2 = 13 - 4l_1^2 - 12l_1l_2 - 9l_2^2.$$

Likewise

$$p_2^2 = OP_2^2 - (3l_1 + 4 \cdot 8l_2)^2 = 32 \cdot 04 - 9l_1^2 - 28 \cdot 8l_1l_2 - 23 \cdot 04l_2^2,$$

and

$$p_3^2 = OP_3^2 - (3 \cdot 8l_1 + 6l_2)^2 = 50 \cdot 44 - 14 \cdot 44l_1^2 - 45 \cdot 6l_1l_2 - 36l_2^2.$$

Hence
$$p_1^2 + p_2^2 + p_3^2 = 95 \cdot 48 - 27 \cdot 44 l_1^2 - 86 \cdot 4 l_1 l_2 - 68 \cdot 04 l_2^2.$$

The problem of choosing l_1 and l_2 to minimize the expression is therefore the problem of choosing them to maximize

$$27 \cdot 44 l_1^2 + 86 \cdot 4 l_1 l_2 + 68 \cdot 04 l_2^2,$$

remembering the restriction that $l_1^2 + l_2^2 = 1$. This is a particular case of Problem 1 of Section 2.9. In this case the problem can be solved by a well-known elementary method. Putting $l_1 = \cos\theta$ and $l_2 = \sin\theta$, θ has to be chosen to maximize

$27 \cdot 44 \cos^2\theta + 86 \cdot 4 \sin\theta \cos\theta + 68 \cdot 04 \sin^2\theta$

$= (47 \cdot 74 - 20 \cdot 3) \cos^2\theta + 86 \cdot 4 \sin\theta \cos\theta + (47 \cdot 74 + 20 \cdot 3) \sin^2\theta$
$= 47 \cdot 74 + 43 \cdot 2 \sin 2\theta - 20 \cdot 3 \cos 2\theta$
$= 47 \cdot 74 - R \cos(2\theta + \alpha)$ \hfill (7.7)

where $R \cos\alpha = 20 \cdot 3$ and $R \sin\alpha = 43 \cdot 2$. This means that

$$\tan\alpha = 43 \cdot 2 / 20 \cdot 3 = 2 \cdot 1281$$

and so $\alpha = 64° \, 50'$.

Expression (7.7) has its maximum value when $2\theta + \alpha = \pi$, and hence when $\theta = 57° \, 35'$. Thus $l_1 : l_2 = 0 \cdot 536 : 0 \cdot 844$.

It is now necessary to attack the general problem. It is convenient to resume notations used both in Sections 7.2 and 7.3, and to assume that t different measurements are made on n different individuals. The results can be tabulated in a matrix $\mathbf{M}$,

$$\mathbf{M} = \begin{bmatrix} x_1^{(1)} & x_2^{(1)} & . & . & x_t^{(1)} \\ x_1^{(2)} & x_2^{(2)} & . & . & x_t^{(2)} \\ . & . & . & . & . \\ x_1^{(n)} & x_2^{(n)} & . & . & x_t^{(n)} \end{bmatrix}.$$

This can be seen as involving n row vectors each with t components, i.e. n vectors corresponding to the n individuals and located in t-space; alternatively it can be seen as t column vectors each with n components, i.e. t vectors corresponding to the t tests and located in n-space. It may be compared with the matrix $\mathbf{X}$ in Section 7.2.

It is convenient to adjust the measurements so that the mean of the n scores on each test is zero, which means that

$$\sum_{r=1}^{n} x_i^{(r)} = 0, \quad i = 1, 2, \ldots, t.$$

It is also possible to normalize so that

$$\sum_{r=1}^{n} x_i^{(r)2} = 1, \quad i = 1, 2, \ldots, t,$$

and the statistician may or may not do this.

We now seek the vector in the t-space which fits the n individual vectors as closely as possible. By 'fits' we mean chosen in the direction which makes the sum of the squares of the perpendicular distances of the n points from it a minimum. Fig. 7.7 will still do if we imagine it extended to t dimensions.

The t measurements on the first individual are contained in the vector $(x_1^{(1)}, x_2^{(1)}, \ldots, x_t^{(1)})$, which is the position vector of point P_1. Using Eq. (2.24) we see that the orthogonal projection of OP_1 on the line l has length equal to $x_1^{(1)} l_1 + \ldots + x_t^{(1)} l_t$. Hence by Pythagoras's theorem, in a space of t dimensions, p_1 the perpendicular distance from P_1 to l is given by

$$p_1^2 = (x_1^{(1)2} + x_2^{(1)2} + \ldots + x_t^{(1)2}) - (x_1^{(1)} l_1 + \ldots + x_t^{(1)} l_t)^2.$$

This expression is more conveniently written using vector notation as

$$p_1^2 = \mathbf{x}^{(1)\prime} \mathbf{x}^{(1)} - (\mathbf{x}^{(1)\prime} \mathbf{l})^2,$$

where
$$\mathbf{l} = (l_1, \ldots, l_t)'.$$

We must remember that $l_1, \ldots, l_t$ are direction cosines and so we have the restriction

$$\mathbf{l}'\mathbf{l} = l_1^2 + l_2^2 + \ldots + l_t^2 = 1. \tag{7.8}$$

To minimize the sum of the p^2's associated with the n points it is necessary to choose l suitably.

This means choosing $\mathbf{l}$ so as to *maximize* the sum

$$\sum_{r=1}^{n} (\mathbf{x}^{(r)\prime} \mathbf{l})^2 = \sum_{r=1}^{n} (x_1^{(r)} l_1 + x_2^{(r)} l_2 + \ldots + x_t^{(r)} l_t)^2 \tag{7.9}$$

subject to condition (7.8).

This enables the initial criterion to be rephrased, and to do so it is appropriate to transfer attention from the t-space to the n-space. Expression (7.9) gives the square of the length of the vector

$$\mathbf{v} = \begin{bmatrix} x_1^{(1)} l_1 + x_2^{(1)} l_2 + \ldots + x_t^{(1)} l_t \\ x_1^{(2)} l_1 + x_2^{(2)} l_2 + \ldots + x_t^{(2)} l_t \\ \vdots \\ x_1^{(n)} l_1 + x_2^{(n)} l_2 + \ldots + x_t^{(n)} l_t \end{bmatrix},$$

this being a vector in the n-space. The problem is now to maximize the length of the vector **v** subject to condition (7.8). Using block matrices **v** may be written

$$\mathbf{v} = [\mathbf{x}_1 \mid \ldots \mid \mathbf{x}_t] \begin{bmatrix} l_1 \\ \cdot \\ \cdot \\ \cdot \\ l_t \end{bmatrix},$$

where $\mathbf{x}_1, \ldots, \mathbf{x}_t$ are the columns of **M**. So $\mathbf{v} = \mathbf{Ml}$, and $\mathbf{v'v} = \mathbf{l'M'Ml}$. Once again we are led to the matrix $\mathbf{M'M}$, and the maximization problem is to maximize $\mathbf{l'M'Ml}$ subject to $\mathbf{l'l} = 1$. This is the familiar ground of Section 2.9, but note that we are once again back in the t-space! The second principal component can now be introduced by seeking to maximize $\mathbf{l'M'Ml}$ by suitable choice from amongst all the vectors orthogonal to the first **l**, just found, and so on for further components. For a fuller discussion of this approach see Reference [7].

The whole of the work just carried out can be given a mechanical interpretation. If we have a solid body, or a set of rigidly connected particles, then we may take co-ordinate axes through the centre of gravity and then seek the axis through the origin about which the moment of inertia of the body is a minimum. The figure corresponding to Fig. 7.7 is now three-dimensional and the expressions calculated as the sums of squares of perpendicular distances are moments of inertia. For details of the theory of moments of inertia a textbook of analytical mechanics should be consulted, but this brief mention of another area of application, at first sight of a very different character, should indicate once again the manifold applicability of the ideas of linear algebra.

References

1. BINFORD, S. R. and L. R., 'Stone Tools and Human Behaviour', *Scient. Am.*, 220, 4, 70–84 (April 1967).
2. BURT, C., *The Factors of the Mind*, Univ. of London Press (1940).
3. COLE, J. P. and KING, C. A. M., *Quantitative Geography*, Wiley (1968).
4. GREGG, J. V., HOSSELL, C. H. and RICHARDSON, J. T., *Mathematical Trend Curves: an Aid to Forecasting*, Oliver and Boyd for ICI (1964).
5. HIGMAN, B., *Applied Group Theoretic and Matrix Methods*, OUP (1955).
6. HOTELLING, H., 'Analysis of a complex of statistical variables in principal components', *J. educ. Psychol.* XXIX, pp. 417–41, 498–520, (1933).
7. KENDALL, M. G., *A Course in Multivariate Analysis*, Griffin (1957).

8. KLAMKIN, M. S., 'On Cooking a Roast', *SIAM Rev.*, 167–9 (April 1961).
9. LOMAX, A., *Folk Song Style and Culture*, Amer. Assn. for Adv. of Sci. (1968).
10. SAWYER, W. W., *The Search for Pattern*, Pelican (1970).
11. SHELDON, W. H., STEVENS, S. S. and TUCKER, W. B., *Varieties of Human Physique*, Harper Brothers (1940).
12. SOKAL, R. R., 'Numerical Taxonomy', *Scient. Am.*, (December 1966).
13. TANNER, J. M., *The Physique of the Olympic Athlete*, Allen and Unwin (1964).
14. THOMSON, G. H., *The Factorial Analysis of Human Ability*, Univ. of London Press (1939).
15. *Children and their Primary Schools*, Vol. 2, Research and Surveys. A report of the Central Advisory Council for Education (England), HMSO (1967).

Solutions and Hints

Chapter 1

Exercise 1.1 Denote the three progressions whose $(r+1)$th terms are $a+rb$, $c+rd$ and $e+rf$ by **x, y** and **z** respectively. Then
$$(de-cf)\mathbf{x}+(af-be)\mathbf{y}+(bc-ad)\mathbf{z}=0.$$
The result follows in general. The exceptional case when all of the coefficients are zero is easily dealt with separately.

Exercise 1.2 Points on the same line through the origin. Addition is represented by the 'parallelogram law'; the two representative points, the origin and the point representing the sum form a parallelogram.

Exercise 1.3 $\mathbf{a}=\mathbf{b}+\mathbf{e}+\mathbf{g}$, $\mathbf{u}=\mathbf{b}-\mathbf{f}$, $\mathbf{v}=\mathbf{d}+\mathbf{f}-\mathbf{b}-\mathbf{e}$, $\mathbf{w}=\mathbf{d}+\mathbf{f}-\mathbf{b}-\mathbf{g}$, $\mathbf{x}=\mathbf{b}-\mathbf{d}$.

$3\mathbf{b}=\mathbf{a}+2\mathbf{u}+\mathbf{v}+\mathbf{w}+2\mathbf{x}$, $3\mathbf{d}=\mathbf{a}+2\mathbf{u}+\mathbf{v}+\mathbf{w}-\mathbf{x}$,
$3\mathbf{e}=\mathbf{a}-\mathbf{u}-2\mathbf{v}+\mathbf{w}-\mathbf{x}$, $3\mathbf{f}=\mathbf{a}-\mathbf{u}+\mathbf{v}+\mathbf{w}+2\mathbf{x}$,
$3\mathbf{g}=\mathbf{a}-\mathbf{u}+\mathbf{v}-2\mathbf{w}-\mathbf{x}$.

No.

Exercise 1.4 The three squares given are linearly independent—look first at the middle term and then at the two diagonals. It is easy to show that given any 3×3 fully magic square the 'magic constant' is necessarily equal to three times the number in the middle of the square. Hence given any magic square we may see how many times it contains the first basic square. Looking at the diagonals determines the multiples it contains of the other two basic squares. The remaining terms then necessarily check.

Exercise 1.5 Calling the first seven of the eight squares listed **a, b, c, d, e, f, g** then the given square is $6\mathbf{a}+\mathbf{b}+8\mathbf{c}+2\mathbf{d}+5\mathbf{e}+2\mathbf{f}+10\mathbf{g}$.

Exercise 1.6 Try to produce Barratt's first square by taking multiples a, b, c, d, e, f, g, h of Botsch's squares.

From the terms in row 2, column 2 $c+f=0$.
From the terms in row 3, column 4 $f+g=0$.
From the terms in row 4, column 3 $a+c=0$.

It follows from these equations that $a+g=0$. But this implies a contradiction in the value of the term in row 1, column 1.

Calling the new squares introduced by Barratt **p, q, r, s**, then
$\mathbf{a}=\mathbf{q}+\mathbf{h}+\mathbf{r}$, $\mathbf{b}=\mathbf{s}+\mathbf{g}+\mathbf{p}$, $\mathbf{c}=\mathbf{p}+\mathbf{e}+\mathbf{q}$, $\mathbf{d}=\mathbf{r}+\mathbf{f}+\mathbf{s}$.

Botsch's other four squares are included in Barratt's. In the same notation Dürer's square is

$$7\mathbf{e} + 10\mathbf{f} + 11\mathbf{g} + 6\mathbf{h} + 3\mathbf{p} + 8\mathbf{q} + 14\mathbf{r} + 9\mathbf{s}.$$

Given any magic square to express in terms of Barratt's eight, first use appropriate multiples of **e, f, g, h** to produce the four centre terms. An appropriate multiple of **p** then gives the correct term in row 1, column 2; of **q** gives the term in row 2, column 4; of **r**, row 4, column 3; and of **s**, row 3, column 1. It can be shown that all of the other terms then necessarily check.

Exercise 1.7 Dimension 6.

Exercise 1.8 The space of $3 \times 3 \times 3$ fully magicubes (i.e. checking by diagonals as well) is of dimension 1. The space of semi-magicubes is of dimension 9.

Exercise 1.10 Subtracting x^r from $(x + 1)^r$ gives a polynomial of degree $r - 1$. Hence, if $f(x)$ is a polynomial of degree r, then

$$f(x + 1) - f(x)$$

is a polynomial of degree $r - 1$. Apply this n times, starting with a polynomial of degree n and this leads to the required result.

Exercise 1.12 Start with the table for $f_k(x)$ and add an extra row above it.

Exercise 1.16 If $px^2 + qxy + ry^2$ is a potential function then it is necessary to have $2p + 2r = 0$, and q may be arbitrary.

Exercise 1.19 (2, 8, 10; 6, 5, 2, 8). (3, 4, 28; 28, 6, 9). (3, 38; 7, 24).

Exercise 1.20 Dimension 6. One suitable basis consists of five triangles like ABC and the pentagon $ABCDE$.

Exercise 1.21 Dimension 4. The relation between the two spaces is discussed further in Exercise 2.23.

Chapter 2

Exercise 2.1 Examples 2.2 and 2.3 produce cycles of period 2.
 Example 2.4 produces a cycle of period 4.
 Example 2.5 has no further effect if it is repeated.

Exercise 2.10 $\mathbf{BA} = \mathbf{I} \Rightarrow \mathbf{BAC} = \mathbf{C}$ and $\mathbf{AC} = \mathbf{I} \Rightarrow \mathbf{BAC} = \mathbf{B}$. Hence $\mathbf{C} = \mathbf{B}$.

Exercise 2.26 Let **H** and **K** be orthogonal matrices and let $\mathbf{L} = \mathbf{HK}$. Then $\mathbf{L'L} = (\mathbf{HK})'\mathbf{HK} = \mathbf{K'H'HK} = \mathbf{K'IK} = \mathbf{I}$.

Exercise 2.27 $\lambda = 1, (2, 1, -2)'; \lambda = -2, (1, 2, 2)'; \lambda = 4, (2, -2, 1)'.$

$$\mathbf{H} = \tfrac{1}{3}\begin{bmatrix} 2 & 1 & 2 \\ 1 & 2 & -2 \\ -2 & 2 & 1 \end{bmatrix}.$$

Hyperboloids of one sheet.

Exercise 2.28 $\lambda = 0, (2, -1, 1)'; \lambda = 3, (1, 1, -1); \lambda = -2, (0, 1, 1)'.$

$$\mathbf{H} = \begin{bmatrix} \dfrac{2}{\sqrt{6}} & \dfrac{1}{\sqrt{3}} & 0 \\ -\dfrac{1}{\sqrt{6}} & \dfrac{1}{\sqrt{3}} & \dfrac{1}{\sqrt{2}} \\ \dfrac{1}{\sqrt{6}} & -\dfrac{1}{\sqrt{3}} & \dfrac{1}{\sqrt{2}} \end{bmatrix}.$$

Hyperbolic cylinders.

Exercise 2.29 $\lambda = 6, (2, 1, 2)'; \lambda = -3, (1, -2, 0)', (4, 2, -5)',$ etc. $\mathbf{H}$ depends on the choice of base vectors for $\lambda = -3$. Hyperboloids of two sheets.

Exercise 2.30

The eigenvector corresponding to $\lambda = 6$ is $(1, 1, -1, 1)'$. Any vector is an eigenvector corresponding to $\lambda = 2$ provided that it satisfies

$$x_1 + x_2 - x_3 + x_4 = 0, \tag{1}$$

that is if it is orthogonal to $(1, 1, -1, 1)'$. There is then a wide choice. One suitable choice is $(1, 1, 1, -1)'$; but $(0, 1, 1, 0)'$ (say) would do just as well. The next eigenvector, which will become a column of $\mathbf{H}$, has to be orthogonal to the vector chosen, so it must satisfy

$$x_1 + x_2 + x_3 - x_4 = 0 \tag{2}$$

as well as Eq. (1). $(1, -1, 1, 1)'$ is a suitable choice, but others are possible. The final choice is then forced to be $(-1, 1, 1, 1)'$. This leads to

$$\mathbf{H} = \tfrac{1}{2}\begin{bmatrix} 1 & 1 & 1 & -1 \\ 1 & 1 & -1 & 1 \\ -1 & 1 & 1 & 1 \\ 1 & -1 & 1 & 1 \end{bmatrix}.$$

If $(0, 1, 1, 0)'$ is chosen at the second stage, at the third stage it is necessary to satisfy $x_2 + x_3 = 0$ as well as Eq. (1). One possible choice now is $(1, 0, 0, -1)'$, although others can be made. The last vector, with this choice is then forced to be $(1, -1, 1, 1)'$. Normalizing appropriately $\mathbf{H}$ is now

$$\tfrac{1}{2}\begin{bmatrix} 1 & 0 & \sqrt{2} & 1 \\ 1 & \sqrt{2} & 0 & -1 \\ -1 & \sqrt{2} & 0 & 1 \\ 1 & 0 & -\sqrt{2} & 1 \end{bmatrix}.$$

SOLUTIONS AND HINTS

Exercise 2.31 Put $\boldsymbol{\xi} = \mathbf{Hx}$. Then $\boldsymbol{\xi}'\boldsymbol{\xi} = \mathbf{x}'\mathbf{H}'\mathbf{Hx} = \mathbf{x}'\mathbf{x}$, since $\mathbf{H}'\mathbf{H} = \mathbf{I}$. Further, if $\boldsymbol{\eta} = \mathbf{Hy}$, and $\mathbf{x}$ and $\mathbf{y}$ are unit vectors then the cosine of the angle between the original vectors is $\mathbf{x}'\mathbf{y}$ and that of the angle between the transformed vectors is $\boldsymbol{\xi}'\boldsymbol{\eta}$. $\boldsymbol{\xi}'\boldsymbol{\eta} = \mathbf{x}'\mathbf{H}'\mathbf{Hy} = \mathbf{x}'\mathbf{y}$. Hence the angle is unaltered.

Exercise 2.34 Put $\mathbf{M}'\mathbf{M} = \mathbf{A}$. Then

$$\mathbf{A}' = (\mathbf{M}'\mathbf{M})' = \mathbf{M}'\mathbf{M}'' = \mathbf{M}'\mathbf{M} = \mathbf{A}.$$

Hence $\mathbf{A}' = \mathbf{M}'\mathbf{M}$ is symmetric.

Since $\mathbf{M}'\mathbf{M}$ is symmetric, as in Theorem 2.4, we may find $\mathbf{H}$ such that $\mathbf{H}'\mathbf{M}'\mathbf{MH} = \text{diag }\{\lambda_1, \ldots, \lambda_n\}$. Hence

$$\mathbf{y}'\mathbf{H}'\mathbf{M}'\mathbf{MHy} = \mathbf{y}'\text{ diag}\{\lambda_1, \ldots, \lambda_n\}\mathbf{y},$$
$$= \lambda_1 y_1^2 + \ldots + \lambda_n y_n^2,$$

this being an identity in the ys.

If any λs are negative it is possible to assign values of the ys such that the right-hand side is negative. But putting $\mathbf{x} = \mathbf{MHy}$ the right-hand side is $\mathbf{x}'\mathbf{x}$ which is the sum of squares. Hence the assumption that any λs are negative leads to a contradiction.

Chapter 3

Exercise 3.1 $\mathbf{G} = \frac{1}{4}\begin{bmatrix} 3 & 2 & 1 \\ 2 & 4 & 2 \\ 1 & 2 & 3 \end{bmatrix}$. $\mathbf{G} = \frac{1}{6}\begin{bmatrix} 11 & 5 & 2 \\ 5 & 5 & 2 \\ 2 & 2 & 2 \end{bmatrix}$.

Exercise 3.2 $G(x, \xi) = -\dfrac{\alpha \xi x}{\alpha + \beta} + x, \qquad 0 \leqslant x \leqslant \xi,$

$$= -\dfrac{\alpha \xi x}{\alpha + \beta} + \xi, \qquad \xi \leqslant x \leqslant 1.$$

Exercise 3.4
$$\mathbf{G} = \tfrac{1}{5}\begin{bmatrix} 4 & 3 & 2 & 1 \\ 3 & 6 & 4 & 2 \\ 2 & 4 & 6 & 3 \\ 1 & 2 & 3 & 4 \end{bmatrix}.$$

Exercise 3.6
$$\mathbf{G} = \tfrac{1}{12}\begin{bmatrix} 25 & 13 & 7 & 3 \\ 13 & 13 & 7 & 3 \\ 7 & 7 & 7 & 3 \\ 3 & 3 & 3 & 3 \end{bmatrix}.$$

Chapter 4

Exercise 4.7
$N = 3.$
$$\mathbf{A} = \begin{bmatrix} 22 & -10 & 0 \\ -10 & 16 & -6 \\ 0 & -6 & 6 \end{bmatrix}.$$

$\lambda = 2, (1, 2, 3)'; \lambda = 12, (1, 1, -1)'; \lambda = 30, (5, -4, 1)'$.

$N = 4$.
$$\mathbf{A} = \begin{bmatrix} 38 & -18 & 0 & 0 \\ -18 & 32 & -14 & 0 \\ 0 & -14 & 22 & -8 \\ 0 & 0 & -8 & 8 \end{bmatrix}.$$

$\lambda = 2, (1, 2, 3, 4)'$; $\lambda = 12, (9, 13, 7, -14)'$; $\lambda = 30, (9, 4, -11, 4)'$; $\lambda = 56, (14, -14, 6, -1)'$.

Exercise 4.8

$$\mathbf{G} = \tfrac{1}{60}\begin{bmatrix} 5 & 5 & 5 \\ 5 & 11 & 11 \\ 5 & 11 & 21 \end{bmatrix}, \quad \mathbf{G} = \tfrac{1}{2520}\begin{bmatrix} 126 & 126 & 126 & 126 \\ 126 & 266 & 266 & 266 \\ 126 & 266 & 446 & 446 \\ 126 & 266 & 446 & 761 \end{bmatrix}.$$

Chapter 5

Exercise 5.1 A verification is provided as follows. Let $\mathbf{T}$ be the matrix required.

Then if $\mathbf{w}$ is in $\mathscr{W}$ we have $\mathbf{Tw} = \mathbf{w}$ and so $\mathbf{TX} = \mathbf{X}$.

Also if $\mathbf{x}$ is orthogonal to $\mathscr{W}$ we have $\mathbf{X}'\mathbf{x} = 0$ and $\mathbf{Tx} = 0$.

If we take it as geometrically obvious that a unique $\mathbf{T}$ is specified by these requirements then we see that $\mathbf{T} = \mathbf{X}(\mathbf{X}'\mathbf{X})^{-1}\mathbf{X}'$ meets them, because $\mathbf{TX} = \mathbf{X}(\mathbf{X}'\mathbf{X})^{-1}\mathbf{X}'\mathbf{X} = \mathbf{X}$, and $\mathbf{Tx} = \mathbf{X}(\mathbf{X}'\mathbf{X})^{-1}\mathbf{X}'\mathbf{x} = 0$.

Exercise 5.3 Using the equations from Exercise 5.2,
$$\mathbf{A} - \lambda_2\mathbf{I} = (\lambda_1 - \lambda_2)\mathbf{A}_1 + (\lambda_3 - \lambda_2)\mathbf{A}_3,$$
and
$$\mathbf{A} - \lambda_3\mathbf{I} = (\lambda_1 - \lambda_3)\mathbf{A}_1 + (\lambda_2 - \lambda_3)\mathbf{A}_2.$$

Multiplying out and using other known relations
$$(\mathbf{A} - \lambda_2\mathbf{I})(\mathbf{A} - \lambda_3\mathbf{I}) = (\lambda_1 - \lambda_2)(\lambda_1 - \lambda_3)\mathbf{A}_1, \quad \text{etc.}$$

Exercise 5.4 $\mathbf{Ax} = \mathbf{x}$ implies that $\mathbf{A}^2\mathbf{x} = \lambda\mathbf{Ax} = \lambda^2\mathbf{x}$, and so $\mathbf{A}^2$ has an eigenvalue λ^2. Similarly $\mathbf{A}^r\mathbf{x} = \lambda^r\mathbf{x}$, and $\mathbf{A}^r$ has an eigenvalue λ^r. Adding appropriate multiples of this equation for various r proves that $f(\mathbf{A})$ has an eigenvalue $f(\lambda)$, and $f(\mathbf{A}) = 0$ implies $f(\lambda) = 0$, since $\mathbf{x} \neq 0$.

For projection matrices $f(\mathbf{A}) = \mathbf{A}^2 - \mathbf{A} = 0$, so the only eigenvalues are roots of $\lambda^2 - \lambda = 0$.

Exercise 5.7 This follows from the orthogonal properties.

Exercise 5.8 G satisfies
$$(pG')' + qG = 0.$$

Multiply by the rth eigenfunction $y_r(x)$, integrate by parts over the intervals (a, ξ) and (ξ, b) and use the boundary conditions. This gives
$$y_r(\xi) = \lambda_r \int_a^b G(x, \xi) y_r(x)\, dx.$$

But this is to say that the Fourier coefficient of G for $y_r(x)$ is $y_r(\xi)/\lambda_r$.

Exercise 5.9 The first relation is proved as in Exercise 5.8, the differential equation for Γ differing from that for G only by the presence of the k^2 term.

For the last part of the question replace $\Gamma(x,\xi)$ under the integral sign by the sum in the first part of the question, interchange integral and summation signs so getting to integrals of the form $\int_a^b y_r(\xi)\, p(\xi)\, d\xi$. By definition this integral is p_r.

Chapter 6

Exercise 6.1
Starting with any pair of integers gives a sequence of approximations to $\sqrt{2}$. $(x,y) \to (x + ky, x + y)$ gives a method for finding $\sqrt{k}$.

Exercise 6.2 The matrix involved is $\begin{bmatrix} 0 & 1 \\ 1 & 0 \end{bmatrix}$. The numbers of H and Cl radicals present interchange at each stage.

Exercise 6.3 Using **u** to denote a column vector composed entirely of ones a matrix **A** has the stochastic property if and only if $\mathbf{Au} = \mathbf{u}$. But this implies that $\mathbf{A^2 u} = \mathbf{Au} = \mathbf{u}$, and so $\mathbf{A}^2$ also has the stochastic property, etc.

Exercise 6.4 There are various ways of drawing up a matrix of unlimited size, using as states all families of the form (0 boys, g girls) and (1 boy, g girls). A simpler model is provided by the matrix

	0 boys	1 boy
0 boys	$1-p$	p
1 boy	0	1

where p is the probability of a birth giving a boy. This neglects the problem of multiple births.

Exercise 6.5 If the probability of red coming up is p, and if we use 'state n' to mean the state occurring after a run of n losses, then the transition matrix is

	0	1	2	3	...
0	p	$1-p$	0	0	...
1	p	0	$1-p$	0	
2	p	0	0	$1-p$	
.	.	.	.	.	

Roulette may be played with or without zero and double zero. This affects the value of p.

Exercise 6.7 Many plausible matrices can be drawn up as the conditions do not specify a unique one. (4, 3) has to be an eigenvector in order to ensure an average of 10 pints per week.

$$\begin{bmatrix} p & 1-p \\ r & 1-r \end{bmatrix}$$

is a suitable matrix provided that $4p + 3r = 4$.

If we make the additional assumption that we do not ever need two pints on consecutive days then we could take $r = 1$ and $p = \frac{1}{4}$.

Exercise 6.8 The players chances are in the ratios $11:8:7$.

Exercise 6.10

$$\mathbf{A}_1 = \tfrac{1}{15}\begin{bmatrix} 6 & 3 & 1 \\ 12 & 6 & 2 \\ 18 & 9 & 3 \end{bmatrix}$$

$$\mathbf{A}_\alpha = \tfrac{1}{30}\begin{bmatrix} 9+3i & -3-6i & -1+3i \\ -12+6i & 9+3i & -2-4i \\ -18-36i & -9+27i & 6-3i \end{bmatrix}$$

and $\mathbf{A}_\beta$ is conjugate to $\mathbf{A}_\alpha$.

$\mathbf{A}_1^n = \mathbf{A}_1$, and $(1, 1, 1)\mathbf{A}^n \sim (1, 1, 1)\mathbf{A}_1 = (\tfrac{12}{5}, \tfrac{6}{5}, \tfrac{2}{5})$.

Exercise 6.11 $(1, \gamma, \gamma)$ and $(\gamma, 1, \gamma)'$. The row vector describes the numbers of radicals present, and so in the long run the radicals are present in the ratios $1:\gamma:\gamma$, whatever the starting ratios.

Exercise 6.12 Construct a sequence of vectors with the property that each is parallel to the normal to an ellipsoid at the point at which the preceding vector was the position vector. This sequence tends to the shortest axis of the ellipsoid.

Chapter 7

Exercise 7.2

$$\mathbf{A} = \begin{bmatrix} -2 & 1 \\ -1 & 1 \\ 0 & 1 \\ 1 & 1 \\ 2 & 1 \end{bmatrix} \text{ and } \mathbf{y} = \begin{bmatrix} y_{-1} \\ y_{-2} \\ y_0 \\ y_1 \\ y_2 \end{bmatrix}.$$

Then
$$\mathbf{A}'\mathbf{A} = \begin{bmatrix} 10 & 0 \\ 0 & 5 \end{bmatrix} \text{ etc.}$$

Exercise 7.4 If the columns of $\mathbf{S}$ are denoted by the column vectors $\mathbf{s}_1, \ldots, \mathbf{s}_f$, then

$$\mathbf{S}'\mathbf{S} = \begin{bmatrix} \mathbf{s}_1' \\ \vdots \\ \mathbf{s}_f' \end{bmatrix} [\mathbf{s}_1 \vdots \ldots \vdots \mathbf{s}_f].$$

Multiplying out the block matrices

$$\mathbf{S'S} = \begin{bmatrix} \mathbf{s}_1'\mathbf{s}_1 & \cdots & \mathbf{s}_1'\mathbf{s}_f \\ \vdots & & \vdots \\ \mathbf{s}_f'\mathbf{s}_1 & \cdots & \mathbf{s}_f'\mathbf{s}_f \end{bmatrix},$$

and the diagonal terms are one, by the normalization, and the other terms are zero by the orthogonality.

To show that $\mathbf{SS'}$ is not necessarily equal to $\mathbf{I}$ it is sufficient to give a counter-example. If $\mathbf{S} = \begin{bmatrix} 1 & 0 \\ 0 & 1 \\ 0 & 0 \end{bmatrix}$ then $\mathbf{S'S} = \mathbf{I}_2$, but $\mathbf{SS'} \neq \mathbf{I}_3$.

Exercise 7.5 $\mathbf{S'S} = \Lambda^{-\frac{1}{2}}\mathbf{H}_f\mathbf{M'MH}_f'\Lambda^{-\frac{1}{2}} = \Lambda^{-\frac{1}{2}}\mathbf{H}_f\mathbf{T'TH}_f'\Lambda^{-\frac{1}{2}}$, since $\mathbf{M'M} = \mathbf{T'T}$.

Hence, using $\mathbf{T} = \Lambda^{\frac{1}{2}}\mathbf{H}_f$, $\mathbf{S'S} = \Lambda^{-\frac{1}{2}}\mathbf{H}_f\mathbf{H}_f'\Lambda^{\frac{1}{2}}\Lambda^{\frac{1}{2}}\mathbf{H}_f\mathbf{H}_f'\Lambda^{-\frac{1}{2}}$.

But $\mathbf{H}_f\mathbf{H}_f' = \mathbf{I}_f$, this being proved as the result in Exercise 7.4, but for rows instead of columns. Hence $\mathbf{S'S}$ reduces to $\mathbf{I}_f$.

Index

abortion, 215
acceleration, 11, 111
angle, 66
angular frequency, 101, 105, 109, 132, 140, 158
area, 11
arithmetical progression, 1
athletics records, 236
axis of symmetry, 75

Barnard, 214
Barratt, 7
base, 38
basis, 2, 5, 7, 16
beam, 139
beetles, 195, 223
bending moment, 141
Bernoulli, 114
Bessel, 117, 160
binary code, 30
Botsch, 6
boundary condition, 22, 111, 115, 119, 123, 139, 141, 148, 150, 157, 174, 186
Bourbaki, 165
bullet, 10

cable nets, 158
chain, 114, 123, 136
characteristic equation, 60, 89, 101, 217, 222, 226
Chebyshev, 171
chemistry, 23, 198, 223, 229, 237
circle, 20
circulant, 156, 226
closure, 2
compound interest, 227
computer, 13, 30, 153, 201, 239, 255
confidential information, 214
conformable, 41

conic, 69, 72
convex, 37
correlation, 214, 247, 251
cosine, 67, 247
curve fitting, 8

degenerate, 51, 66, 92, 158
degrees of freedom, 97
denumerability, 37
determinant, 46, 59, 75, 218
difference equation, 21, 147, 180
differences, 14
differential equation, 21, 113, 121, 138, 140, 161, 174, 176, 228, 229
dimension, 2, 3, 5, 6, 8, 13, 22, 36, 39, 106, 243
dimensional analysis, 23
direction cosine, 68, 78
direction ratio, 68, 72
dominant eigenvalue, 221
drum heads, 160
Dürer, 7

economics, 215, 229
ectomorph, 253
education, 201, 239
educational testing, 244
eigenfunction, 113, 120, 139, 141, 163, 176, 183, 190
eigensequence, 150
eigenspace, 65, 66
eigenvalue, 58, 63, 78, 81, 85, 93, 101, 113, 117, 120, 134, 139, 142, 149, 156, 159, 166, 174, 179, 190, 194, 197, 217, 220, 229, 243, 250
eigenvector, 58, 63, 75, 78, 81, 85, 101, 113, 117, 120, 134, 139, 142, 149, 156, 159, 166, 174, 194, 197, 217, 220, 229
electrical filters, 159

271

elementary step, 48
elementary transformation, 54
ellipse, 69, 71, 90, 94, 233, 243
ellipsoid, 69, 75, 93, 95
endomorph, 253
energy, 108, 127, 131, 180
enlargement, 71
ergodic chain, 206
error correction, 31, 69
Euler, 214
exponential function, 229
extrapolation, 9

factor, 245, 250, 255
Fibonacci, 20
field, 27, 28, 39
flow, 25
flow chart, 52, 61
folk song, 255
Fourier, 37, 167
   coefficient, 169, 180, 182, 186
   series, 167, 182
function space, 8, 21

Galois, 58
geography, 134, 250
geometric sequence, 20
GF(2), 28, 35
golden number, 223
Gram, 69
Green, 121, 139, 144, 150, 159, 186
group, 35
growth, 227

half-turn, 43
Hamming, 30
heat conduction, 174
Hilbert, 69, 192
homology, 35
hyperbola, 69, 71
hyperboloid, 75

independence, 6, 7, 36, 39, 51, 65, 96
infinite series, 116, 161, 229
inner product, 41, 66, 176
integral equation, 190
interpolation, 8
invariant, 43, 49

Kepler, 24
Kirchhoff, 32
Klamkin, 236

Lagrange, 9, 97, 167
Legendre, 139
Leontief, 217
line of best fit, 232, 257
linear dependence, 6, 7, 25, 26, 28, 30, 36, 57, 65, 96
linear manifold, 37, 38
linear space, 1, 5, 21, 37
Liouville, 173, 178, 190
Lomax, 255

magic square, 3
Manchester, 252
mapping, 85, 89, 93
Markov chain, 194, 206, 217
Mathematics Teaching, 20
matrix, 40, 53, 75
   banded, 148, 181
   block, 208, 224
   circulant, 156, 226
   connectivity, 134
   correlation, 232
   covariance, 232
   diagonal, 80, 83, 84, 87, 249
   flexibility, 107, 156, 180
   inverse, 43, 46, 56, 80, 131, 151
   normal, 221
   orthogonal, 80, 84, 247
   projection, 165, 167
   stiffness, 105, 153, 155, 180
   stochastic, 204, 206, 214, 217
   sub-, 208
   symmetric, 66, 71, 81, 85, 93, 108, 182
   transformation, 54
   transpose, 57
   unit, 41
maximim, 85, 92, 172, 259
McDougall, 224
Mersenne, 41
mesomorph, 253
minimum, 85, 92, 172, 234, 258
module, 27
moment of inertia, 259

networks, 32, 68, 134
Nim, 28
normal, 72, 74
normal matrix, 221
normal co-ordinates, 106, 129
normal mode, 102, 109, 112, 116, 127, 130, 141, 148, 163, 192
normalization, 68, 166, 169, 245

INDEX

numbers
  Fibonacci, 20
  tetrahedral, 18
  triangular, 18

Olympic athletes, 254
operator
  central difference, 149, 159
  differenetial, 112, 115, 126, 131, 138, 145, 163, 176, 188
  integral, 189
  inverse, 112, 120, 124, 126, 145, 159
  self-adjoint, 176, 180
  symmetric, 173, 178, 179, 188
orthogonal, 68, 75, 80, 84, 169, 176, 184, 219, 221

parabola, 69
parallelogram, 70
partial differential equation, 110, 161
particles, 98, 130, 147
Pascal, 16
Peaker, 240
pendulum, 99
pentagram, 7
physics, 23
planetary motion, 24
Plowden Report, 239, 252
Poisson, 124
Polya, 19
polynomial, 14, 146
  Chebyshev, 171
  interpolation, 8
  Legendre, 139
potential, 22, 124
power series, 116, 229
principal axis, 75
principal component, 241, 249, 256
probability, 204, 210, 213
projection, 67, 77, 165, 257
Pythagoras, 66, 69, 185, 259

quadratic, 8, 74
quadric, 69, 72, 75, 78, 83, 92

rank, 51, 95, 246
Rayleigh, 88, 105, 127, 132, 146, 177, 192
reaction, 198
Redfern, 201
reflection, 43
regression, 235, 237, 249
resonance, 126, 186, 222
ring, 27
rocket, 9
roots of unity, 157
rotation, 90, 252

Sawyer, 14, 19, 236
scalar multiplication, 2, 3, 5, 27, 36, 58, 63, 75, 112, 176
Schmidt, 69
self-adjoint, 176, 180, 182
semi-magic square, 3
sequence, 36, 195, 200
series
  Bessel, 117, 160
  Fourier, 167, 182
shear, 45, 90
shearing force, 141
Sheldon, 252
Simpson, 11
simultaneous equations, 46, 228
singular, 51
Snakes and Ladders, 211
space
  function, 8, 21, 112, 115, 132, 176
  linear, 1, 2, 5, 21, 97, 179
  sub-, 6, 22, 166
  vector, 2, 5, 37, 38, 96, 147
spectral decomposition, 221, 223, 247
square
  magic, 3
  root, 194, 224
  semi-magic, 3
standard book number, 68
statistics, 232
stochastic, 194
stretch, 44, 90, 91
string, 23, 110, 118, 121, 148, 152, 170, 173, 185
Sturm, 173, 178, 190
summation, 17
symmetric
  matrix, 66
  operator, 173, 176, 179

tangent, 71
Tanner, 254
Theon of Smyrna, 194
ticker tape, 10
torus, 35
transformation, 41, 77, 80, 84
  inverse, 43, 79, 92
  orthogonal, 75
  row, 54

274 INDEX

transition probabilities, 205, 210
turkey, 235

United States of America, 250

variance, 251
vector, 2, 5, 24, 26, 28, 30, 33, 36, 38, 49, 96, 176, 258

base, 3, 27, 36, 169
bound, 70
free, 70
normalized, 68, 166, 169, 246
orthogonal, 68, 76, 246
unit, 67

wave, 140
whirling, 143